Emerging Technologies for Promoting Food Security

Related Titles

Environmental Assessment and Management in the Food Industry
(ISBN 978-1-84569-552-1)

Handbook of Water and Energy Management in Food Processing
(ISBN 978-1-84569-195-0)

Handbook of Organic Food Safety and Quality
(ISBN 978-1-84569-010-6)

Emerging Technologies for Promoting Food Security

Overcoming the World Food Crisis

Edited by

Chandra Madramootoo

ELSEVIER

AMSTERDAM • BOSTON • CAMBRIDGE • HEIDELBERG
LONDON • NEW YORK • OXFORD • PARIS • SAN DIEGO
SAN FRANCISCO • SINGAPORE • SYDNEY • TOKYO
Woodhead Publishing is an imprint of Elsevier

WP WOODHEAD PUBLISHING

Woodhead Publishing is an imprint of Elsevier
80 High Street, Sawston, Cambridge, CB22 3HJ, UK
225 Wyman Street, Waltham, MA 02451, USA
Langford Lane, Kidlington, OX5 1GB, UK

ISBN: 978-1-78242-335-5

British Library Cataloguing-in-Publication Data
A catalogue record for this book is available from the British Library

Library of Congress Cataloging-in-Publication Data
A catalog record for this book is available from the Library of Congress

Publisher: Nikki Levy
Acquisition Editor: Rob Sykes
Editorial Project Manager: Josh Bennett
Production Project Manager: Lisa M. Jones
Designer: Greg Harris

Typeset by TNQ Books and Journals
www.tnq.co.in

Contents

List of Contributors

Stuart W. Bunting Bunting Aquaculture, Agriculture and Aquatic Resources Conservation Services, Glemsford, Suffolk, UK

H.-X. Chang Department of Crop Sciences, University of Illinois, Urbana, IL, USA

Marie-Josée Dumont Bioresource Engineering Department, McGill University, Ste-Anne-de-Bellevue, QC, Canada

Helen Fyles Faculty of Agriculture and Environmental Sciences, Macdonald Campus, McGill University, Ste-Anne-de-Bellevue, QC, Canada

G.L. Hartman USDA-Agricultural Research Service, University of Illinois, Urbana, IL, USA; Department of Crop Sciences, University of Illinois, Urbana, IL, USA

C.B. Hill Agricen Sciences, LLC, Pilot Point, TX, USA

Simerjeet Kaur Department of Plant Science, McGill University, Ste-Anne-de-Bellevue, QC, Canada

David C. Little Institute of Aquaculture, University of Stirling, Stirling, UK

Chandra Madramootoo Faculty of Agriculture and Environmental Sciences, Macdonald Campus, McGill University, Ste-Anne-de-Bellevue, QC, Canada

Haritika Majithia Department of Plant Science, McGill University, Ste-Anne-de-Bellevue, QC, Canada

Valérie Orsat Bioresource Engineering Department, McGill University, Ste-Anne-de-Bellevue, QC, Canada

M.L. Pawlowski Department of Crop Sciences, University of Illinois, Urbana, IL, USA

Vijaya Raghavan Bioresource Engineering Department, McGill University, Ste-Anne-de-Bellevue, QC, Canada

Jaswinder Singh Department of Plant Science, McGill University, Ste-Anne-de-Bellevue, QC, Canada

Kevin M. Wade Department of Animal Science, McGill University, QC, Canada

Key Drivers of Food Insecurity

1

Helen Fyles, Chandra Madramootoo
Faculty of Agriculture and Environmental Sciences, Macdonald Campus, McGill University, Ste-Anne-de-Bellevue, QC, Canada

1. Introduction

Global food insecurity affects one in eight people or 870 million worldwide, with two-thirds of these people found in Sub-Saharan Africa, India, and China (FAO, 2013). Food insecurity is driven by multiple factors that may affect food production, availability, and stability of food supplies, as well as food prices at national, community, or household levels. The key drivers of global food insecurity are rising demand for food, climate change, natural resource availability, biofuel production, and a lack of public and private investment in infrastructure (roads, electricity, etc.), and research and development in the agricultural sector. The price of food also plays a major role in food insecurity and is directly and indirectly affected by the key drivers. The degree of importance of each key driver varies between countries or regions according to their unique set of physical, economic, and social circumstances.

2. Food Prices

When food prices rise, the world's poorest households, which spend a large share of their income on food, are forced to reduce the quality and the quantity of food consumed. This leads to increased food insecurity at the household level (Brinkman et al., 2010). The harmful effect of sharply higher food prices at the country level is in proportion to the level of net food imports, which is currently a concern for the majority of developing countries (Valdéz and Foster, 2012). Wheat consumption has risen throughout Africa in the past two decades, and 90% of this increase has been provided by imports. This has raised the dependence of many countries on global wheat supplies and increased their vulnerability to international prices (IDRC, 2013). Food prices are directly related to agricultural input costs (fertilizers, pesticides, fuel for machinery and irrigation, transporting products to market, and food processing), which in turn are related to the price of fossil fuels. The food sector currently accounts for about 30% of the world's total energy consumption, and cereal prices tend to follow oil prices (FAO et al., 2011). Actual or perceived worldwide grain shortages caused by adverse weather conditions also have an immediate effect on international prices.

Emerging Technologies for Promoting Food Security. http://dx.doi.org/10.1016/B978-1-78242-335-5.00001-9

3. Rising Demand for Food

3.1 Population Growth

Population growth drives the demand for food. Over the last 20 years, the world population has risen by 1.6 billion, and 78% of this increase was in low-income, food-deficit countries (Mazzocchi et al., 2012). The world's population reached 7.2 billion in August 2013 and is projected to reach over 9.6 billion by 2050 (UN, 2013). For each year until 2050, farmers will try to feed an additional 40–86 million people. Amplifying this demand for food are the middle-income countries, where over 70% of the world's population live and where the rapid growth in living standards and dietary diversity are putting increasing pressure on world food security (World Bank, 2013a).

Due to the demographic transition away from rural areas, over 50% of the world's population currently lives in urban areas, and almost all urban population growth in the next 30 years will occur in cities of developing countries (WHO, 2014). This shift in population creates new pressures on global food security, and increased urbanization may result in the loss of productive land surrounding cities. In Indonesia, for example, about 40,000–50,000 ha per year of agricultural land is taken out of production due to urbanization (Schultz, 2011). Peri-urban areas are becoming increasingly important sources of food, but where and how this food is produced is complex (Lerner and Eakin, 2011). In order to achieve food security, a significant change may be required from smallholder farming towards food production for the urban population. This implies an increase in farm size and production of higher-value crops on land surrounding urban areas (Schultz, 2011).

3.2 Changes in Food Consumption Choices

Rising incomes, rapid urbanization, and globalization have led to improved access to a greater variety of foods and resulted in changes to food consumption patterns over the last 30 years. Changes in per capita gross national incomes (Table 1) illustrate the substantial increase in the spending power of millions of people in the middle-income countries.

Table 1 Changes in Gross National Income (GNI)

Country/Income Level	Gross National Income (GNI) per Capita (current US $) 1980	2012	Change in GNI 1980–2012 (%)
Low income	302	1187	293
Middle income	650	4369	572
High income	8764	37,595	329
India	270	1530	466
China	220	5740	2500

Source: World DataBank. World Development Indicators, http://databank.worldbank.org/data.

China in particular has seen huge gains in per capita income, which have allowed more people to expand their diet to include significant amounts of expensive items (meats, milk, and eggs), which require large amounts of feed crops to produce. Since 1980, the amount of per capita daily energy (kilocalories) coming from animal products (milk, meat, eggs) has risen by almost 300% in China, compared to only a 20% rise for vegetal products (Table 2). India saw smaller gains in energy from animal products (80%), but because these two countries account for such a large proportion of the world's population, even small dietary shifts can have very large effects on world agricultural crop demand. The increasing share of animal products in the diet is part of a dietary transition that also includes a higher intake of fats, fish, vegetables, and fruit at the expense of cereals and tubers.

If meat and dairy consumption continue to climb at a similar rate to the past, by 2050, the world will require an estimated additional 522 million tons of grain livestock feed. This accounts for approximately 40% of the grain produced worldwide (Ahuja, 2013; AAAP, 2012). How the increased demand for livestock products will be met—through diversion of grain crops or grazing—will have a significant influence on global food security. The average efficiency of converting feed into meat or milk is currently considerably lower than what is possible (AAAP, 2012). In addition, ruminants produce food from land otherwise unsuitable for growing crops and use plant resources that do not compete with human food. The success of future research in this field will have a great influence on how dietary shifts drive global food insecurity.

4. Climate Change

Climate change drives food security at both the global scale, where it impacts the production of internationally traded grains, and at regional or local scales, where people, often the poor and food insecure, depend heavily on local food production (Lobell and Gourdji, 2012). Higher concentrations of CO_2, widespread temperature increases, rising sea levels, shrinking glaciers, shifting precipitation patterns, and greater frequency

Table 2 Food Supply from Vegetal and Animal Products

	Vegetal Products (kcal/capita/day)			Animal Products (kcal/capita/day)		
Country/Region	**1980**	**2009**	**% Increase**	**1980**	**2009**	**% Increase**
Africa	2029	2353	16	192	207	8
China	1989	2342	18	174	694	299
India	1875	2113	13	116	209	80
North America	2191	2658	21	970	1001	3
World	2103	2330	11	390	501	28

Source: FAOSTAT Food Balance Sheets. Updated June 2012.

and duration of extreme weather events (droughts, floods, heat stress) are all associated with climate change (IPCC, 2007). There is a wide range of predicted impacts on crop production due to regional variations in growing season length, minimum and maximum temperatures, frequency and intensity of precipitation, and the spread of crop pests and diseases. Climate change impacts on crop yields will also vary widely with soil texture, nutrient and organic matter levels, and the ability of farmers to mitigate precipitation and temperature changes through agrochemical use, irrigation, drainage, use of different crop varieties, and altering farm management techniques such as timing of field operations and use of conservation agriculture.

4.1 Variations in Precipitation

Expected declines in crop productivity due to climate change often coincide with countries that currently have a high level of food insecurity (Wheeler and von Braun, 2013). More frequent and severe droughts are likely to exacerbate water scarcity in many parts of the world, and Sub-Saharan Africa, where 97% of the agricultural production is rain-fed, is particularly vulnerable to the consequences of changes in precipitation patterns and temperature (World Bank, 2013b). Across Africa, serious declines are expected in maize, wheat, sorghum, millet, groundnuts, and cassava production, as well as a steep decline in the extent of savanna grasslands, leading to reduced forage available for grazing animals. This will have a very negative impact on regional food security.

In South Asia, the region with the world's largest concentration of poor people, the most important crops are rice and wheat, accounting for about 50% and 40% of crop production, respectively (World Bank, 2013b). Drought and extreme rainfall over the last 50 years in India have already reduced rain-fed rice yields by about 6%, and wheat yields have not increased in 10 years (Auffhammer et al., 2012; Lobell et al., 2012). Increasingly variable monsoon rainfall necessary to replenish groundwater in South Asia, coupled with reduced flows from glacial melting, could have very negative consequences to food security in South Asia (World Bank, 2013b). An estimation of crop production changes in South Asia to 2050 under climate change suggest that daily per capita food availability will drop from 2424 to 2241 kcal (Nelson et al., 2010).

The world's major grain exporters—the United States, Canada, France, Australia, Argentina, and the Russian Federation—produce the majority of their grain in highly productive rain-fed conditions. Declines in water availability are projected to affect France, Australia, and parts of the Russian Federation (IPCC, 2007). This, coupled with increased heat stress described above, suggests global grain production will become more variable and unpredictable.

4.2 Predictions of Crop Yields

Projections of future crop productivity under climate change are generated by general circulation and crop growth models and are subject to uncertainty due to the necessity of models to oversimplify real-world complexity (Müller et al., 2011). Projected impacts are a balance between crop yield increases due to elevated CO_2 and decreases due to rising average growing season temperatures, more frequent and higher extreme temperatures, and changes to precipitation patterns, all of which depend to some extent on the choice of crop growth and climate model (Knox et al., 2012). Pest and disease occurrence, reduction in glacial water and depletion of groundwater supply as an irrigation source, sea level rise, flooding, and the influence of modern inputs and management practices are often not accounted for in model predictions (Piao et al., 2010). With these uncertainties in mind, it is not surprising that there are wide variations in predictions of climate change effects on major grain yields and regional crop production effects (Table 3). In general, modeling results for a range of sites find that in mid- to high-latitude regions, rising CO_2, moderate to medium local increases in temperature (1–3 °C), and rainfall changes, can have small beneficial impacts on crop yields (Easterling et al., 2007). In low-latitude regions, even moderate temperature increases (1–2 °C) are likely to have negative yield impacts for major cereals. In Africa, a review of model predictions suggest yield reductions of up to 40% across crop types (wheat, maize, sorghum, millet) and subregions by 2050 and beyond (Knox et al., 2012).

4.3 Effects of Rising CO_2

Wheat, rice, soybeans (C3 crops) and maize (C4 crop) provide about 75% of the calories consumed worldwide either directly or indirectly (Cassman, 1999). C3 crops generally produce 20–30% more above-ground dry matter as CO_2 levels rise to 550–580 ppm (expected by the late twenty-first century), even as temperatures rise to heat stress levels (Wang et al., 2012). However, recent work by the U.S. Department of Agriculture (Ainsworth, 2013) suggests that C3 crop responses to rising CO_2 may be less optimistic and more complicated than previously thought; in a cooler-than-average growing season, soybean yield increased with CO_2 levels as temperatures warmed, but in a warmer-than-average growing season, warming caused a significant reduction in seed yield (−44%), even with elevated levels of CO_2. C4 crops (which also include sugarcane and sorghum) do not respond to changes in CO_2 at ambient temperatures, although they do have about a 10% higher productivity as temperatures rise (Wang et al., 2012). High levels of CO_2 also cause stomata to partially close, which improves water use efficiency in both C3 and C4 crops. However, the interaction between high CO_2 concentrations and soil moisture is not yet understood. For example, the response of soybean yield to elevated CO_2 and drought stress varies between growing seasons, and elevated CO_2 does not always compensate for the deleterious effects of drought (Ainsworth, 2013). Valerio et al. (2011) note that a C3 crop does well against a C4 weed at high CO_2 levels only when soil moisture is adequate.

4.4 Effects of Changing Growing Seasons and Higher Temperatures

Over the last 30 years in the United States (contiguous 48 states), growing season length has increased by about 10 days (EPA, 2013). Climate change is expected to continue to result in longer growing seasons in northern regions (IPCC, 2007).

Table 3 **Model Predictions of Climate Change Effects on Various Crops**

Grain	Region	Model Predictions (2030–2090)	References
WHEAT	Worldwide	5.4% Decrease in wheat yield per 1 °C increase in temperature.	Lobell and Field (2007)
	China	Rain-fed yield reduced in northern regions and increased in southern regions.	Lv (2013)
		Average rain-fed wheat production will decline by 11%.	Ju et al. (2013)
	USA	National wheat production expected to increase.	Izaurralde et al. (2003)
	Europe	Increase in the northern or cooler areas from 2% to 30%; decrease in southern areas.	IPCC (2007) and Olesen et al. (2011)
	India–Pakistan	Up to 51% of high-yield areas might become heat-stressed, resulting in significant yield losses.	Ortiz et al. (2008)
		A 2–5% decrease in yield potential for temperature rise of 0.5–1.5 °C in India.	Cruz et al. (2007)
		In rice-wheat cropping systems in central India, wheat yields may decline by 4–41% depending on planting date.	Jalota et al. (2012)
	Australia	Nationally, crop yields drop slightly.	Hennessy et al. (2007)
	Africa	Decline of 17%.	Knox et al. (2012)
RICE	Worldwide	Increases in air temperatures above current levels will reduce rice yield by about 10% per 1 °C rise.	Hatfield et al. (2011)
	China	Decline of 4–12%.	Ju et al. (2013)
		Increase of 7%.	Xiong et al. (2008)
	Asia	+0.5 to −1.5%.	Masutomi et al. (2009)
	India	In rice-wheat cropping systems in central India, rice yields may vary by −29 to +15%, depending on planting date and CO_2 fertilization effect.	Jalota et al. (2012)
MAIZE	Worldwide	Decline in yields by 14%.	Lobell and Burk (2010)
	China	Dryland yields to decrease 14–22%.	Ju et al. (2013)
	USA	Decline in rain-fed production by 2–8%.	Hatfield et al. (2011)
		Irrigated production to increase.	Izaurralde (2003)
	Europe	Northern areas: 30–50% increase in area suitable for production.	IPCC (2007)
		Southern regions: large decline.	Audsley et al. (2006)
	India	2–5% Decrease in yield potential for temperature rise of 0.5–1.5 °C.	Cruz et al. (2007)
	South Asia	−16%.	Knox et al. (2012)
	Africa	−5%.	Knox et al. (2012)
SOYBEANS	USA	Upper Midwest yield may increase. Southern states yields are expected to decline by 2.4% for 0.8 °C increase.	Hatfield et al. (2011)
	Europe	Northern regions: increased viability.	Audsley et al. (2006)
		Southern regions: large decline.	

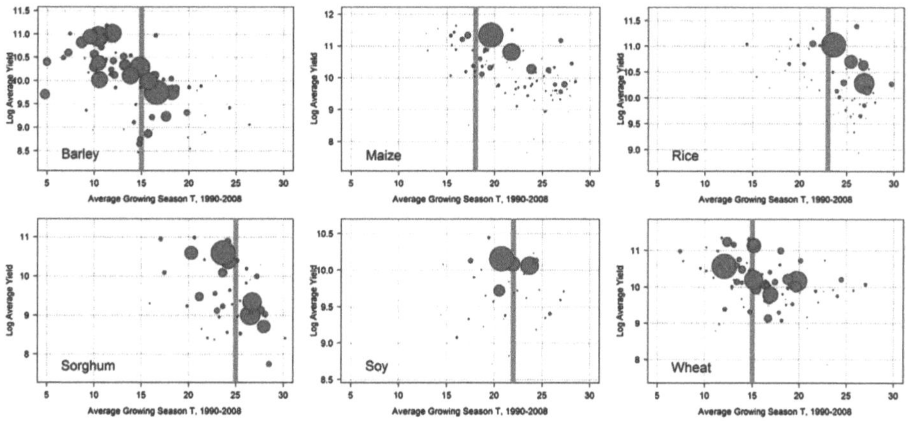

Figure 1 Average yields for six major crops plotted against average growing season temperature (T). Each dot represents a single country, with the size of the dot proportional to total national production for that crop. Gray vertical lines indicate optimal temperature for yields based on experiments, as reported by Hatfield et al. (2011). The highest national yields are typically observed close to the optimum temperature, with lower average yields for warmer countries. Also apparent is that many countries that are major producers are currently above optimal temperature.

A longer growing season can allow farmers to diversify crops or have multiple harvests, but it can also change the types of crops grown, encourage invasive species, pests, or weed growth, or increase demand for irrigation.

When current average growing season temperatures of the world's major crop production areas are compared to optimum growing season temperatures for each crop, it is evident that most of the world's maize and rice, and significant portions of wheat, soybeans, sorghum, and barley, are growing in regions where temperatures are already above optimum (Figure 1) (Lobell and Gourdji, 2012). Although many factors affect yields other than temperature, crops growing above their optimum would likely be very sensitive to further increases in temperature. In addition, in temperate regions, rising temperatures will likely improve the winter survival of many pests and enable them to extend their range (Jarvis et al., 2010). It is also likely that there will be more frequent peaks of high temperatures in these regions, which can drastically reduce yields if they occur during critical crop-development stages. Rice grain yield is reduced about 10% per 1 °C temperature increase above 25 °C, until reaching zero yield at 35–36 °C (Luo, 2011). Temperatures above 30 °C reduce maize grain yield, and the maximum temperature for grain filling in wheat is about 35 °C. In the Canadian prairies, 35 °C is rarely exceeded and currently occurs 0.5 days per growing season. However, this is expected to increase to 4.4 days per season for 2040–2069 (Gameda and Qian, 2011). In Africa, each 1 °C increase in mean temperature (between 1961 and 2010) during the maize growing season resulted in yields reduced by over 10% in eight countries and 5–10% in 10 countries, although in four relatively cool countries, yields increased by more than 5% (Shi and Tao, 2014).

Important grain-growing regions in China, the United States, Russia, and Canada are all likely to be affected by increased heat stress, although sowing date, genotype choice, and irrigation may help to reduce the impact on yields (Teixeira et al., 2013).

5. Availability of Natural Resources

Fresh water scarcity, land degradation, and soil nutrient and organic matter loss are all on the rise and reduce the quality and quantity of natural resources available for food production.

5.1 Water Availability

Agriculture accounts for 70% of global freshwater withdrawals worldwide and more than 90% of its consumptive use, although there is wide regional variation (Hoekstra et al., 2012). Northern Africa uses 77% of its freshwater resources in irrigation, and the Middle East 40%, while Latin America, North America, and Europe use less than 2% (Frenken and Gillet, 2012). Over the last century, water use has been growing globally at more than twice the rate of population increase, and competition for fresh water resources is increasing between agricultural, industrial, domestic, and ecological uses (Palaniappan and Gleick, 2009). As populations grow and diets continue to shift towards meat, milk, sugar, oils, and vegetables—which typically require more water to produce than cereals—it is expected that 17% more water will be required by 2020 to meet food production needs (Palaniappan and Gleick, 2009). Currently, 40% of the world's food production (including 43% of its cereal production) is grown on land equipped for irrigation (Siebert and Döll, 2010).

Water scarcity is already a problem in many parts of the world, and an increasing number of regions are reaching the limit at which reliable water services can be delivered. An estimated 2.7 billion people are living in river basins that experience severe water scarcity during at least one month of the year, and almost half a billion people experience severe water scarcity for at least six months of the year (Hoekstra et al., 2012).

Irrigation uses water from renewable sources (surface or groundwater that is renewed annually) and/or nonrenewable sources (groundwater bodies that have a negligible rate of recharge). The contribution of nonrenewable groundwater to meet irrigation needs has been increasing rapidly worldwide, and in 2000 it was estimated that nearly 20% of the global irrigation demand was met by nonrenewable groundwater abstraction (Wada et al., 2012). This means that about 8% of the world's food supply is currently produced using unsustainable water sources. Groundwater abstraction augments surface water supply for large-scale irrigation of crops, particularly rice and wheat, in the Indus, Ganges, Yangtze, and Yellow river basins at rates that generally exceed both natural and return-flow recharge (Immerzeel and Bierkens, 2012). Rapidly growing populations in these regions means that more food needs to be produced, but current overuse of nonrenewable supplies indicates that in the long term, there will be insufficient quantities of fresh water to sustain food security.

One particularly vulnerable basin is the Indo-Gangetic basin (IGB), home to a rapidly growing population of almost a billion people and producing most of the wheat and half the rice grown in India and Pakistan (Sharma et al., 2010). Agriculture accounts for 91% of the water use in the IGB, and total water demand already exceeds supply. Extensive use of groundwater for irrigation has resulted in a rapid decline in water table depths, in many areas by more than 4 m in a 10-year period (Gopalakrishnan, 2011). Glacier melt from the Himalayas is a major water source for the IGB, particularly the Indus Basin, and the combination of decreasing ice mass, early snowmelt, and increased winter stream flow under climate change will negatively impact groundwater recharge and food production. Overuse of nonrenewable groundwater is also taking place on the North China Plain (NCP), where more than 30% of China's crops are produced (Jiao, 2010). The NCP went from 1800 powered wells in the 1960s to more than 700,000 wells by 2000, which resulted in a rapid lowering of the groundwater table by as much as 1 m each year, and many wells are now hundreds of meters deep. Water samples drawn from the aquifer under the NCP were estimated to be as old as 30,000 years, suggesting that groundwater recharge times are slow (Jiao, 2010). In the United States, the world's top exporter of wheat, the Ogallala Aquifer provides water for about 25% of the irrigated cropland in the United States and supports nearly 20% of the wheat, corn, cotton, and cattle produced nationally (McGuire, 2009). Declining aquifer levels and concerns that groundwater quality will degrade in the future suggest that this aquifer cannot sustain current water use in the long term (McGuire, 2009; Scanlon et al., 2010). Ongoing reliance on nonrenewable groundwater for crop production is a threat to global food security.

5.2 Contamination of Freshwater

Salinization and contamination of freshwater caused by inappropriate fertilizer use, poorly managed irrigation, and pollution and seawater intrusion of coastal aquifers due to over pumping and sea level rise further reduce available water resources. Worldwide, FAO estimates that 34 Mha of irrigated land (11% of total irrigated area) are affected by some level of salinity, with Pakistan, China, the United States, and India accounting for 21 Mha (FAO, 2011). Increased use of nitrogen (N) and phosphorus (P) fertilizers has caused widespread contamination of surface and groundwater.

5.3 Loss of Cropland and Land Degradation

Cropland is increasingly being converted to other uses due to urbanization, industrialization, and population growth. Urbanization is predicted to cause the loss of 1.6–3.3 Mha annually of prime agricultural land over the next 20 years (Lambin and Meyfroidt, 2011).

Deforestation to clear large tracts of land for agricultural production and grazing is often the initial cause of soil degradation. Estimated using satellite data, 1.5 million km^2 of land was deforested between 2000 and 2012, with the majority of this loss in the tropics (Hansen et al., 2013). Some of this deforestation occurs as small farmers are compelled to clear and cultivate marginal lands on steep hills. Without proper management, deforestation reduces soil water-holding capacity, encourages surface water runoff and soil erosion, and decreases groundwater recharge. Long-term intensive soil cultivation and overgrazing, particularly on dry or sloping soils, can lead to erosion and loss of organic matter and nutrient-rich topsoil, resulting in a decline in soil structure and aggregation and reduced water and nutrient holding capacity. In addition, fertilizer applications in excess of crop requirements can lead to pollution of water sources.

Degradation of cropland appears to be most prevalent in Africa, affecting 65% of cropland areas, compared with 51% in Latin America and 38% in Asia (Oldeman, 1991). Degradation of pasture is also most serious in Africa, affecting 31% of pastureland, compared with 20% in Asia and 14% in Latin America. Wood et al. (2000) estimated that 40% of the world's agricultural land is moderately degraded, and a further 9% strongly degraded, reducing global crop yield by 13%.

5.3.1 Soil Erosion

Each year, millions of hectares of cropland are abandoned due to low productivity caused by soil erosion. In developing countries, soil erosion is particularly severe on small farms that are often located on marginal lands where the soil quality is poor and topography frequently steep. Global estimations of soil erosion vary from 25 to 45 billion tons of soil removed from agricultural lands per year (Quinton et al., 2010). Soil loss rates are estimated to average 30–40 tons/ha/year across Africa, Asia, and South America, and go as high as 1000 tons/ha/year in vulnerable areas in China (Ananda and Herath, 2003). Past soil erosion for the African continent as a whole has caused an average annual crop yield decline of 8.2% and 6.2% for Sub-Saharan Africa (Lal, 1995). Soil erosion also causes siltation of reservoirs and waterways, and agrochemical runoff and subsequent water pollution.

5.3.2 Soil Nutrient Loss

Nutrient removal from agricultural soils through erosion, leaching, and crop uptake needs to be balanced through external inputs, and although N and P are often replenished, other nutrients are not. In many Asian and Latin American countries, current rates of soil nutrient depletion are unsustainable. In Africa, nutrient deficiencies are considered the major biophysical factor limiting small-scale farm production (reviewed by Bossio et al., 2010). In Sub-Saharan Africa, nutrient depletion, soil erosion, and desertification cause annual yield reductions ranging from 2% to 40%, with a mean loss of 8.2% for the region (Jones et al., 2013). In India, soil nutrient deficiencies (S, K, or micronutrients) have significantly reduced the effectiveness of N and P fertilizers and now limit crop yields; the amount of grain grown per kg of nutrient applied declined from 13.4 kg of grain in 1970 to 3.7 kg of grain (reviewed by Jones et al., 2013). Loss of nutrients through erosion also wastes expensive nutrients added as fertilizer. Worldwide, it is estimated that erosion removes 23–42 million tons of N and 12–22 million tons of inorganic P from agricultural fields each year (Quinton et al., 2010).

5.3.3 Soil Salinization

Salinization due to poor irrigation practices is a serious threat to sustainable irrigated agricultural production. Globally, an estimated 950 M ha of salt-affected lands occur in arid and semi-arid regions, and roughly 20% of irrigated land suffers from secondary salinization and waterlogging (Wood et al., 2000). Roughly 250,000 to 500,000 ha worldwide are taken out of production every year due to salinization (Nellemann et al., 2009).

6. Markets and Trade

Trade and markets are essential but complex tools used to balance unequal global and regional distribution of supply and demand of food. At the international level, trade policies, agricultural support policies in developed countries, and market liberalization have led to increased food imports by developing countries and left them vulnerable to food price surges (Madramootoo and Fyles, 2012). When food prices rose sharply in 2007–2008, many countries implemented trade measures including export taxation, quantitative restrictions on exports, and reductions in import tariffs in an attempt to safeguard domestic food security. Ultimately, these measures drove food prices higher and exacerbated global food security (Rutten et al., 2013).

At regional and local levels, inadequate policies, institutions, and rural infrastructure have led to agricultural markets that do not function efficiently. Agriculture plays a predominant role in the livelihoods of most people in developing countries, especially in Sub-Saharan Africa and South Asia, and any strategy for improving food security must include rapid growth in the agricultural sector. However, poorly functioning markets, weak domestic consumer demand, and lack of export possibilities are major constraints on agricultural growth prospects for many developing countries. Without access to appropriate markets, small farmers pay more for inputs (seeds, fertilizer, pesticides) and receive less for their produce (von Grebmer et al., 2013). In order for this situation to be reversed, trade policies, market access, and availability of credit and insurance must be adjusted at multiple levels.

With current growth in high-value agricultural trade, consumer demand and regulations related to sanitary, phytosanitary, and quality issues play an increasingly important role in determining agricultural trade opportunities (von Grebmer et al., 2013). Understanding the economics of such nontariff barriers and providing a dispute settlement process to resolve trade conflicts is a critical part of designing effective policies that allow the agricultural sectors of developing countries to bring their products to market.

7. Investments in the Agricultural Sector

National and international investments in agriculture and food production have declined since the 1980s, due to a perception that the world's food supply was secure and that countries were becoming more self-sufficient in food. There was also a focus on exporting high-value agricultural products rather than on food production in many developing countries. Between 1979 and 2004, funding of Official Development Assistance (ODA) to agriculture declined from 18% to 3.5% of the total ODA from the OECD countries (World Bank, 2008). It also declined in absolute terms, from a high of about $8 billion (in 2004, US dollars) in 1984 to $3.4 billion in 2004. Public agricultural research and development expenditures around the world also slowed significantly after the 1970s. In much of Sub-Saharan Africa, the public sector spent less on agricultural R&D in 2000 than it had 10 years earlier (Beintema and Stads, 2010). Declining investments in the agricultural sector meant that Ministries of Agriculture were cut back, extension services diminished, and research and innovation underfunded. There was also a lack of funding for infrastructure in rural areas including roads, electricity, irrigation, water storage, and an absence of rural markets, collection and certification centers, and transport for perishables, all of which have impeded food production gains and limited improvements to food security.

Farmers and other private businesses will only invest in agriculture if expected returns compensate for perceived risks and provide higher returns than other investments. To encourage these private investments, governments and development agencies need to reduce risks and provide incentives through agricultural research appropriate to the domestic agricultural sector and creation of good rural infrastructure, a robust input supply, output-processing facilities, and transparent market institutions (FAO, 2012). Appropriate public investment is therefore critical to farmer investments, which in turn are necessary if economic growth, poverty reduction, and food and nutrition security are to occur.

Land-based investments ("land grabs") are becoming increasingly important to governments in food-importing countries, which are ensuring long-term food security in their own countries through land acquisition overseas. Depending on how they are structured, such investment may be beneficial to smallholder farmers, or they may carry environmental and social risks that fall disproportionally on local people (Cotula et al., 2011).

8. Biofuel Production

Climate change, an urgent need to reduce CO_2 emissions, declining stocks of oil and gas, and issues of national energy security are driving biofuel production throughout the world. The emerging biofuels market is now a significant source of demand for agricultural commodities such as sugar, maize, and oilseeds. These crops, predominantly used as food in the past, are now also used as feedstock for producing ethanol and biodiesel. This production of crops for biofuels diverts land and water away from food production, reduces food availability, and drives up food prices. After public policies to support the biofuels sector were initiated, global biofuel production increased from 17.8 to 105.6 billion liters between 2000 and 2012 (REN21, 2013). A slowdown in biofuel production since 2010 is in response to higher feedstock prices, lower profit margins, and concerns about competition with food production for land and water resources and the long-term sustainability of production. Even so, 52 countries worldwide (including 25 high-income; 23 middle-income; and 4

Table 4 **Amount of Crops Used Worldwide for Biofuel Production**

	Amount Used for Biofuels	Proportion of Global Crop Production
Crop	Million Tons	%
Sugarcane	293	16
Maize[a]	127	14
Soybeans	19.9	8
Rapeseed	15.5	6
Sugar beets	10.8	4
Wheat[b]	7.4	1

[a]Includes 2 million tons used by Canada to produce ethanol.
[b]Includes 0.7 million tons used by Canada to produce ethanol (Dessureault, 2011).

low-income countries) have biofuel obligations or blend mandates, which continue to drive demand for ethanol and biodiesel (REN21, 2013).

Production of biofuels now uses a significant proportion of global crop production (Table 4). Maize and sugarcane used as biofuel feedstocks represent 16% and 14% of global production, respectively, and current world market prices of these crops are three times higher than they were in 2000. It should be noted that large quantities of meal, produced as a byproduct of biofuel crops, are used as feed in the livestock industry (particularly pork and chicken). Part of the profitability of biofuel production comes from the sale of these by-products, and biofuel production, animal feed prices, and livestock production are closely linked.

The world's largest ethanol producer is the United States, and it is estimated that 117 million tons of maize were used to produce ethanol in 2012, about 40% of the total maize grown in the country and about 3% of the total farmland (RFA, 2013; USDA, 2012). Subsidies and incentives have mostly focused on advanced biofuels (those that do not use food grains) and the Renewable Fuel Standard in the United States required that the volume of renewable fuel blended into transportation fuel be 63 billion liters in 2013 and reach 136 billion liters by 2022. In 2013, 25% of the renewable fuel was mandated to come from advanced biofuel or biomass-biodiesel. Since biodiesel production currently represents less than 8% of the total and advanced biofuel production is negligible, the 2013 target was not met, and maize-based ethanol continues to make up the majority of renewable fuel consumed in the United States.

The European Union (EU) is the world's major producer of biodiesel, and its production required 15.5 M tons of rapeseed and 6.5 M tons of soybeans in 2012. This represents 80% of total EU rapeseed production, while most of the soybeans are imported from South America and the United States (Krautgartner et al., 2012).

EU member states are mandated to reach a minimum of 10% for renewable energy consumed in transport in 2020. In 2012, the European Commission published a proposal to limit the use of food-based biofuels to 5% (or half of the 10% minimum) by 2020. It also suggested an enhanced incentive scheme to promote the use of advanced biofuels from feedstocks that would reduce the use of food crops used for biofuels.

9. Conclusions

Expanding populations and growing economic well-being in emerging economies and changes in urban demographics are increasing the global demand for food. The increasing popularity of meat and dairy products require ever larger amounts of feed to produce and push the demand for grain crops higher. As global demand for food rises, climate change, natural resource degradation, and biofuel production make meeting this demand difficult. Increasing temperatures, water scarcity and drought, changes to precipitation patterns, more extreme weather events, reduction in glacial water, sea level rise, and flooding are expected to have negative effects on crop production worldwide, although the effects of climate change on food crops of local importance (for example root crops, vegetables, pulses), which are often critical to local diets, has not been well studied. Climate change impacts, increasing water scarcity, and loss of productive land due to poor soil management can be addressed, at least to some extent, in the agricultural sector through technological improvements and investments in infrastructure, research, and capacity building.

Building up reserves of grain when harvests are plentiful to prepare for times when there is scarcity is also a useful tool to prevent food shortages and price volatility and lessen food insecurity. Toward the end of 2010, the countries of the Association of Southeast Asian Nations, plus China, Japan, and South Korea (altogether known as ASEAN+3) established an emergency rice reserve to help ensure long-term food security in the region (Briones, 2011). Global information systems are improving access to information on weather, production, and stocks around the world, and this will enhance the ability of organizations such as the Food and Agriculture Organization of the United Nations to improve crop supply forecasts and better anticipate and prepare for grain shortages and price spikes (Wright, 2009).

References

Ahuja, V., 2013. Asian livestock: challenges, opportunities and the response. In: Proc. International Policy Forum Held in Bangkok, Thailand, 16–17 August 2012. Animal Production and Health Commission for Asia and the Pacific, International Livestock Research Institute and Food and Agriculture Organization of the United Nations.

Ainsworth, E.A., 2013. How will the two major midwest crops respond to global climate change? Results from ten years of investigation at Soyface 2013. In: Paper at Water, Food, Energy for a Sustainable World. 3–6 November, Tampa, Florida. https://scisoc.confex.com/crops/2013am/webprogram/Paper78729.html.

Ananda, J., Herath, G., 2003. Soil erosion in developing countries: a socio-economic appraisal. Journal of Environmental Management 68, 343–353.

Asian-Australian Association of Animal Production Societies (AAAP), 2012. Symposium on optimizing feed use efficiency in ruminant animal production systems. In: AAAP ANIMAL SCIENCE CONGRESS Held in Bangkok from 26 to 30 November 2012. Reported by FAO. http://www.fao.org/ag/againfo/home/en/news_archive/2013_FAO_Symposium_highlights_importance_enhancing_feed_conversion_efficiency.html. [August 2013].

Audsley, E., Pearn, K.R., Simota, C., Cojocaru, G., Koutsidou, E., Rounsevell, M.D.A., Trnka, M., Alexandrov, V., 2006. What can scenario modelling tell us about future European scale agricultural land use, and what not? Environment Science and Policy 9, 148–168.

Auffhammer, M., Ramanathan, V., Vincent, J.R., 2012. Climate change, the monsoon, and rice yield in India. Climatic Change 111, 411–424.

Beintema, N.M., Stads, G.J., 2010. Public Agricultural R&D Investments and Capacities in Developing Countries: Recent Evidence for 2000 and beyond. International Food Policy Research Institute (IFPRI).

Bossio, D., Geheb, K., Critchley, W., 2010. Managing water for managing land: addressing land degradation to improve water productivity and rural livelihoods. Agricultural Water Management 97, 536–542.

Brinkman, H.J., de Pee, S., Sanogo, I., Subran, L., Bloem, M.W., 2010. High food prices and the global financial crisis have reduced access to nutritious food and worsened nutritional status and health. Journal of Nutrition 140, 153S–161S.

Briones, R.M., 2011. Regional Cooperation for Food Security: The Case of Emergency Rice Reserves in the ASEAN Plus Three. Asia Development Bank No. 18, November 2011.

Cassman, K.G., 1999. Ecological intensification of cereal production systems: yield potential, soil quality, and precision agriculture. Proceedings of National Academy of Sciences 96, 5952–5959.

Cotula, L., Vermeulen, S., Mathieu, P., Toulmin, C., 2011. Agricultural investment and international land deals: evidence from a multi-country study in Africa. Food Security 3, 99–113.

Cruz, R.V., et al., 2007. Asia. Climate change 2007: impacts, adaptation and vulnerability. In: Parry, M.L., Canziani, O.F., Palutikof, J.P., van der Linden, P.J., Hanson, C.E. (Eds.), Contribution of Working Group II to the Fourth Assessment Report of the Intergovernmental Panel on Climate Change, Cambridge University, Cambridge, UK, pp. 469–506.

Dessureault, D., 2011. Canada Biofuels Annual. USDA Foreign Agricultural Service. Global Agricultural Information Network. GAIN Report Number CA11036.

Easterling, W.E., Aggarwal, P.K., Batima, P., et al., 2007. Food, fibre and forest products. In: Parry, M.L., Canziani, O.F., Palutikof, J.P., Van Der Linden, P.J., Hanson, C.E. (Eds.), Climate Change 2007: Impacts, Adaptation and Vulnerability. Contribution of Working Group II to the Fourth Assessment Report of the Intergovernmental Panel on Climate Change. Cambridge University Press, Cambridge, pp. 273–333.

Environmental Protection Agency (EPA), 2013. Climate Change Indicators in the United States: Length of Growing Season. http://www.epa.gov/climatechange/science/indicators/ [August 2013].

FAO, 2013. FAO Statistical Yearbook. World Food and Agriculture. http://www.fao.org/docrep/018/i3107e/i3107e00.htm. [August 2013].

FAO, 2012. State of Food and Agriculture. Investing in Agriculture for a Better Future. FAO, Rome. http://www.fao.org/docrep/017/i3028e/i3028e.pdf.

FAO, 2011. The State of the World's Land and Water Resources for Food and Agriculture (SOLAW) – Managing Systems at Risk. Food and Agriculture Organization of the United Nations, Rome and Earthscan, London.

FAO, International Fund for Agricultural Development, International Monetary Fund, Organisation for Economic Co-operation and Development, United Nations Conference on Trade and Development, World Food Programme, World Bank, World Trade Organization, International Food Policy Research Institute, United Nations High-Level Task Force on the Global Food Security Crisis, 2011. Price Volatility in Food and Agricultural Markets: Policy Responses. Policy Report to the G20, 3 May.

Frenken, K., Gillet, V., 2012. Irrigation Water Requirement and Water Withdrawal by Country. Aquastat November 2012 http://www.fao.org/nr/water/aquastat/water_use_agr/index.stm [September 2013].

Gameda, S., Qian, B., 2011. Climate change and Canadian agriculture. In: Madramootoo, C.A. (Ed.), Water Management for Global Food Security. McGill Institute for Global Food Security, McGill University, Montreal, Quebec. 122 pp.

Gopalakrishnan, M., 2011. An overview of India's water management scenario for ensuring food security. In: Madramootoo, C.A. (Ed.), Water Management for Global Food Security. McGill Institute for Global Food Security, McGill University. PQ.122 pp.

von Grebmer, K., Headey, D., Béné, C., Haddad, L., Olofi nbiyi, T., Wiesmann, D., Fritschel, H., Yin, S., Yohannes, Y., Foley, C., von Oppeln, C., Iseli, B., 2013. 2013 Global Hunger Index: The Challenge of Hunger: Building Resilience to Achieve Food and Nutrition Security. Welthungerhilfe, International Food Policy Research Institute, and Concern Worldwide, Bonn, Washington, DC, and Dublin.

Hansen, M.C., Potapov, P.V., Moore, R., et al., 2013. High-resolution global maps of 21st-century forest cover change. Science 342, 850–853.

Hatfield, J.L., et al., 2011. Climate impacts on agriculture: implications for crop production. Agronomy Journal 103, 351–370.

Hennessy, K., Fitzharris, B., Bates, B., Harvey, N., Howden, S.M., Hughes, L., Salinger, J., Warrick, R., 2007. Australia and New Zealand. Climate change 2007: impacts, adaptation and vulnerability. In: Parry, M.L., Canziani, O.F., Palutikof, J.P., van der Linden, P.J., Hanson, C.E. (Eds.), Contribution of Working Group II to the Fourth Assessment Report of the Intergovernmental Panel on Climate Change. Cambridge University Press, Cambridge, UK, pp. 507–540.

Hoekstra, A.Y., Mekonnen, M.M., Chapagain, A.K., Mathews, R.E., Richter, B.D., 2012. Global monthly water scarcity: blue water footprints versus blue water availability. PLoS One 7, e32688. http://dx.doi.org/10.1371/journal.pone.0032688.

Immerzeel, W.W., Bierkens, M.F.P., 2012. Asia's water balance. Nature Geoscience 5, 841–884.

International Research Development Council (IDRC), 2013. Facts and Figures on Food Biodiversity. http://www.idrc.ca/EN/Resources/Publications/Pages/ArticleDetails.aspx?-PublicationID=565 [September 2013].

IPCC, 2007. Climate change 2007: impacts, adaptation and vulnerability. In: Parry, M.L., Canziani, O.F., Palutikof, J.P., van der Linden, P.J., Hanson, C.E. (Eds.), Contribution of Working Group II to the Fourth Assessment. Report of the Intergovernmental Panel on Climate Change. Cambridge University Press, Cambridge, UK. 976 pp.

Izaurralde, R.C., Rosenberg, N.J., Brown, R.A., Thomson, A.M., 2003. Integrated assessment of Hadley Center (HadCM2) climate-change impacts on agricultural productivity and irrigation water supply in the conterminous United States: Part II. Regional agricultural production in 2030 and 2095. Agricultural and Forest Meteorology 117, 97–122.

Jalota, S.K., Kaur, H., Ray, S.S., Tripathi, R., Vashisht, B.B., Bal, S.K., 2012. Mitigating future climate change effects by shifting planting dates of crops in rice–wheat cropping system. Regional Environmental Change 12, 913–922.

Jarvis, A., Ramirez, J., Anderson, B., Leibig, C., Aggarwal, P., 2010. Scenarios of climate change within the context of agriculture. In: Reynolds, M.P. (Ed.), Climate change and crop production, vol. 1. Cabi.

Jiao, L., 2010. Water shortages loom as Northern China's aquifers are sucked dry. Science 328, 1462–1463.

Jones, D.L., Cross, P., Withers, P.J.A., DeLuca, T.H., Robinson, D.A., Quilliam, R.S., Harris, I.M., Chadwick, D.R., Edwards-Jones, G., 2013. Nutrient stripping: the global disparity between food security and soil nutrient stocks. Journal of Applied Ecology 50, 851–862.

Ju, H., van der Velde, M., Lin, E., Xiong, W., Li, Y., 2013. The impacts of climate change on agricultural production systems in China. Climatic Change 1–12.
Knox, J., Hess, T., Daccache, A., Wheeler, T., 2012. Climate change impacts on crop productivity in Africa and South Asia. Environmental Research Letters 7, 034032.
Krautgartner, R., Henard, M.-C., Rehder, L., Boshnakova, M., Dobrescu, M., Flach, B., Wilson, J., Wideback, A., Bettini, O., the Group of FAS Oilseeds Specialists in the EU, 2012. EU-27 Oilseeds and Products Annual. USDA Foreign Agricultural Service. Global Agricultural Information Network. GAIN Report Number E70016.
Lal, R., 1995. Erosion-crop productivity relationships for soils of Africa. Soil Science Society of America Journal 59, 661–667.
Lambin, E.F., Meyfroidt, P., 2011. Global land use change, economic globalization, and the looming land scarcity. Proceedings of the National Academy of Sciences 108 (9), 3465–3472.
Lerner, A., Eakin, H., 2011. An obsolete dichotomy? Rethinking the rural–urban interface in terms of food security and production in the global south. The Geographical Journal 177, 311–320.
Lobell, D.B., Gourdji, S.M., 2012. The influence of climate change on global crop productivity. Plant Physiology 160 (4), 1686–1697.
Lobell, D.B., Burk, M., 2010. Economic impacts of climate change on agriculture to 2030. In: Reynolds, M.P. (Ed.), Climate Change and Crop Production, vol. 1. Cabi.
Lobell, D.B., Field, C.B., 2007. Global scale climate-crop yield relationships and the impact of recent warming. Environment Research Letters 2, 1–7.
Lobell, D.B., Sibley, A., Ortiz-Monasterio, J.I., 2012. Extreme heat effects on wheat senescence in India. Nature Climate Change 2, 186–189.
Luo, Q., 2011. Temperature thresholds and crop production: a review. Climatic Change 109, 3–4.
Lv, Z., Liu, X., Cao, W., Zhu, Y., 2013. Climate change impacts on regional winter wheat production in main wheat production regions of China. Agricultural and forest meteorology 171, 234–248.
Madramootoo, C., Fyles, H., 2012. Synthesis of findings from the four McGill conferences on global food security 2008–2011. Food Security 4, 307–317.
Masutomi, Y., Takahashi, K., Harasawa, H., Matsuoka, Y., 2009. Impact assessment of climate change on rice production in Asia in comprehensive consideration of process/parameter uncertainty in general circulation models. Agriculture, Ecosystems and Environment 131, 281–291.
Mazzocchi, M., Shankar, B., Traill, B., 2012. The Development of Global Diets since 1992: Influences of Agri-food Sector Trends and Policies. FAO Commodity and Trade Policy Research Working Paper No. 34.
McGuire, V.L., 2009. Water-Level Changes in the High Plains Aquifer, Predevelopment to 2007, 2005–06, and 2006–07. U.S. Geological Survey Scientific Investigations Report 2009–5019, 9 pp. Available at: http://pubs.usgs.gov/sir/2009/5019/.
Müller, C., Cramer, W., Hare, W.L., Lotze-Campen, H., 2011. Climate change risks for African agriculture. Proceedings of the National Academy of Sciences of the United States of America 108, 4313–4315.
Nelson, G.C., Rosegrant, M.W., Koo, J., Robertson, R., Sulser, T., Zhu, T., Ringler, C., et al., 2010. The Costs of Agricultural Adaptation to Climate Change. Washington, DC.
Nellemann, C., MacDevette, M., Manders, T., Eickhout, B., Svihus, B., Prins, A.G., Kaltenborn, B.P. (Eds.), 2009. The Environmental Food Crisis – the Environment's Role in Averting Future Food Crises. A UNEP Rapid Response Assessment. United Nations Environment Programme, GRID-Arendal. www.grida.no.
Working Paper and Reprint 88/4. In: Oldeman, L.R. (Ed.), 1991. Guidelines for General Assessment of the Status of Human-induced Soil Degradation. International Soil Reference and Information Centre, Wageningen.

Olesen, J.E., Trnka, M., Kersebaum, K.C., Skjelvåg, A.O., Seguin, B., Peltonen-Sainio, P., Rossi, F., Kozyra, J., Micale, F., 2011. Impacts and adaptations of European crop production systems to climate change. Eur J Agron 34, 96–112.

Ortiz, R., Sayre, K.D., Govaerts, B., Gupta, R., Subbarao, G.V., Ban, T., Hodson, D., Dixon, J.M., Ortiz-Monasterioa, J.I., Reynolds, M., 2008. Climate change: can wheat beat the heat? Agriculture Ecosystems Environment 126, 46–58.

Palaniappan, M., Gleick, P., 2009. Peak water. In: Gleick, P.H. (Ed.), The World's Water 2008-2009. The Biennial Report on Freshwater Resources. Pacific Institute for Studies in Development, Environment, and Security.

Piao, S., Ciais, P., Huang, Y., Shen, Z., Peng, S., Li, J., et al., 2010. The impacts of climate change on water resources and agriculture in China. Nature 467 (7311), 43–51.

Quinton, J., Govers, G., Van Oost, K., Bardgett, R., 2010. The impact of agricultural soil erosion on biogeochemical cycling. Nature Geoscience 3, 311–314.

REN21, 2013. Renewables 2013 Global Status Report. REN21 Secretariat, Paris.

Renewable Fuels Association, 2013. Ethanol Industry Outlook. Battling for the Barrel. http://www.ethanolrfa.org/pages/annual-industry-outlook.

Rutten, M., Shutes, L., Meijerink, G., 2013. Sit down at the ball game: how trade barriers make the world less food secure. Food Policy 38, 1–10.

Scanlon, B.R., Gates, J.B., Reedy, R.C., Jackson, W., Bordovsky, J., 2010. Effects of irrigated agroecosystems: 2. Quality of soil water and groundwater in the southern High Plains, Texas. Water Resources Research 46 (9).

Schultz, B., 2011. Role of tidal lowlands for food production in the humid tropics. In: Madramootoo, C.A. (Ed.), Water Management for Global Food Security. McGill Institute for Global Food Security. McGill University, Montreal, Quebec. 122 pp.

Sharma, B., Amarasinghe, U., Xueliang, C., de Condappa, D., Shah, T., Mukherji, A., Bharati, L., Ambili, G., Qureshi, A., Pant, D., Xenarios, S., Singh, R., Smakhtin, V., 2010. The Indus and the Ganges: river basins under extreme pressure. Water International 35, 493–552.

Shi, W., Tao, F., 2014. Vulnerability of African maize yield to climate change and variability during 1961–2010. Food Security 6, 471–481.

Siebert, S., Döll, P., 2010. Quantifying blue and green virtual water contents in global crop production as well as potential production losses without irrigation. Journal of Hydrology 384, 198–217.

Teixeira, E.I., Fischer, G., van Velthuizen, H., Walter, C., Ewert, F., 2013. Global hot-spots of heat stress on agricultural crops due to climate change. Agricultural and Forest Meteorology 170, 206–215.

United Nations, Department of Economic and Social Affairs, Population Division, 2013. World Population Prospects: The 2012 Revision, Highlights and Advance Tables. Working Paper No. ESA/P/WP.228.

United States Department of Agriculture (USDA), 2012. Ogallala Aquifer Initiative. Available at: http://www.nrcs.usda.gov/Internet/FSE_DOCUMENTS/stelprdb1048828.pdf [September 2013].

Valdés, A., Foster, W., 2012. Net food-importing developing countries: who they are, and policy options for global price volatility. In: ICTSD Programme on Agricultural Trade and Sustainable Development; Issue Paper No. 43. International Centre for Trade and Sustainable Development, Geneva, Switzerland. www.ictsd.org.

Valerio, et al., 2011. Quantifying the effect of drought on carbon dioxide-induced changes in competition between a C3 crop (tomato) and a C4 week (*Amaranthus retroflexus*). Weed Research 51, 591–600.

Wada, Y., Beek, L., Bierkens, M., 2012. Nonsustainable groundwater sustaining irrigation: a global assessment. Water Resources Research 48 (6).
Wang, et al., 2012. A meta-analysis of plant physiological and growth responses to elevated temperatures. Oecologia 169, 1–13.
Wheeler, T., von Braun, J., 2013. Climate change impacts on global food security. Science 341, 508.
WHO, 2014. Urban Population Growth. Global Health Observatory. http://www.who.int/gho/urban_health/situation_trends/urban_population_growth_text/en/.
Wood, S., Sebastian, K., Scherr, S., 2000. Pilot Analysis of Global Ecosystems: Agroecosystems. International Food Policy Research Institute and World Resources Institute, Washington, DC.
World Bank, 2013a. Middle Income Countries. http://www.worldbank.org/en/country/mic [February 2014].
World Bank, 2013b. Turn Down the Heat: Climate Extremes, Regional Impacts, and the Case for Resilience. A report for the World Bank by the Potsdam Institute for Climate Impact Research and Climate Analytics. World Bank, Washington, DC.
Wright, B.D., 2009. International Grain Reserves and Other Instruments to Address Volatility in Grain Markets. World Bank Policy Research Working Paper Series, 5028 http://ssrn.com/abstract=1471126.
World Bank, 2008. World Development Report. Agriculture for Development. econ.worldbank.org.
Xiong, W., et al., 2008. Future cereal production in China: modelling the interaction of climate change, water availability and socio-economic scenarios. In: The Impacts of Climate Change on Chinese Agriculture—Phase II Final Report (AEA Group, 2008).

Part One

Emerging Technologies for Plant and Animal Food Security

Emerging Genetic Technologies for Improving the Security of Food Crops

2

Jaswinder Singh, Simerjeet Kaur, Haritika Majithia
Department of Plant Science, McGill University, Ste-Anne-de-Bellevue, QC, Canada

1. Introduction

Food security is a growing global concern. More than one billion people are undernourished, and five million children under the age of five die annually of hunger (Schulman, 2012). The world population will double by 2050 at the current rate, and food production will have to be doubled in less than 60 years while simultaneously alleviating poverty that makes the increasingly expensive food unavailable to so many. Continuous constraints on natural resources because of such a population explosion makes increasing crop yield by bringing more land into cultivation not a realistic option (Wall and Smit, 2005; Godfray et al., 2010).

The increasing demand for food and energy in a rapidly expanding world population presents serious concern. Providing these resources in an environmentally sustainable manner with diminishing usable land and water is the greatest challenge. In this changing scenario, modern genomic and genetic engineering tools, in conjunction with plant breeding, can address these challenges. Plant breeders must make use of latest technologies that may increase the selection efficiency of desirable plants. Modern molecular and genomics approaches such as marker-assisted selection are powerful tools to help select precisely the crops with beneficial traits. Above all, the methods used to maintain sustainable agriculture should be environmentally friendly and beneficial for consumption.

Progress in conventional plant breeding has become rather slow since the mid-1990s. In the last 20 years, we have witnessed stagnation in the yield increase of several important crops such as rice, wheat, and maize. Legume production also did not increase much per unit area in the last few years. Another way to overcome this yield barrier is to put greater effort toward protecting the crops against yield losses. Making the crops more resilient to different stress factors can also assuage losses (Khera et al., 2013).

The genetic backgrounds of plants along with favorable environmental factors lead to successful crop production. The arrangement of different genes in a crop variety play a critical role in how the crop will perform in terms of yield, pest and disease resistance, drought and flood tolerance, fertilizer use efficiency, and its nutritional worth. Plant breeders are continually trying to produce the best crop by utilizing an optimum combination of genes, which can be beneficial for future use and growing conditions. Recent availability of high-throughput and cost-effective genome sequencing and -omics technologies can be integrated to fulfill the future needs of

Emerging Technologies for Promoting Food Security. http://dx.doi.org/10.1016/B978-1-78242-335-5.00002-0

an expanding population. This chapter focuses on the main challenges and emerging genetic technologies to cope with food security concerns, especially through the development of value-added enhanced crop plants.

1.1 Competition for Land, Water, and Energy

1.1.1 Land Shortage

When densely populated countries start to industrialize, demand for grain increases as incomes rise; the cultivable area decreases, leading to the fall of grain production. This negative correlation results in higher grain import. Countries that were self-sufficient soon become up to 70% dependent on other countries, Japan being the best example (Chang et al., 2013). In a country that is undergoing rapid industrialization, initially rising incomes lead to increased grain consumption directly, and eventually indirectly via livestock production. As countries industrialize, agricultural land starts being used more and more for industrial and residential purposes, construction of roads and highways, etc. If the leftover farming land is too scarce or fragmented to yield valuable returns, farmers mostly just abandon their plots and seek employment elsewhere. Moreover, industrialization demands labor away from farms toward cities, resulting in less double cropping—where one crop is harvested and replaced quickly by the next crop. Also, an increase in incomes leads to diversified diets of the consumers, creating more demand for exotic fruits and vegetables—which shifts the farmers' focus to growing high-value crops instead of grains (Tscharntke et al., 2012).

Using crops for nonfood uses, such as animal fodder, biofuels, and other industrial products, reduces the already low amount of food available to the people. Globally, only 62% of the cultivated crop is used for human consumption, and 35% is used as animal feed. Three percent of the agricultural produce is allocated for bioenergy and other industrial products (Foley et al., 2011).

1.1.2 Water Shortage

Seventy percent of the world's freshwater is utilized for irrigation. Underground water is being pumped out from irrigation wells in several countries at a faster rate than it can be replenished by rainwater. Water tables are drastically falling in densely populated countries like India and China for this reason. Because most of the shallow aquifers have dried up in China, the Chinese wheat crop production has declined by 8% since 1997. More than half of India's hand-dug wells have already dried up. Electricity shortages are becoming endemic as most of the resources are being used to pump water from as deep as 1 km. A little less than 200 million Indians consume grains that are irrigated from well water. This poses dire consequences to the survival of millions of Indians (Brown, 2009).

1.1.3 Energy

The global food system today is completely dependent on fossil fuels such as oil and natural gas (Jones, 2001). The energy prices have increased three times over the past eight years, and so has the cost of fertilizers. All of this has resulted in an increased cost of foods. Decreasing supplies of fossil fuels have opened an opportunity for using agricultural crops for biofuels. As a result, a significant amount of land is now cultivating crops such as sugar canes and grains to produce ethanol and biodiesel. This vicious circle of producing more biofuels by reducing food production is not able to prevent hunger and starvation among the poor (Runge and Senauer, 2007).

1.2 Biotic and Abiotic Stresses

Crops are endlessly facing numerous challenges due to climatic temperature fluctuations or in the form of insect attack, pathogen infection, and drought. Pests and diseases cause serious crop yield losses, up to 20%, every year due to their constantly evolving nature. Pest-related losses alone contribute to an average of 14–25% of the total global agricultural production (DeVilliers and Hoisington, 2011). Every year, 26% of soybeans, 28% of wheat, 31% of maize, 37% of rice, and 40% of potatoes are lost due to pests (Oerke, 2006; Oerke and Dehne, 2004). In cotton, chickpeas, pigeon peas, tomatoes, and sunflowers, however, crop losses from pests are in the range of 15–30%. Moreover, losses due to pathogens, animal pests, and weeds are around 16%, 18%, and 34%, respectively (Dhaliwal et al., 2010). To increase yield in abiotic stress conditions such as drought and salinity, agriculture must be focused toward selecting crops that grow well under optimal conditions, as they would be more likely to also grow well under stress conditions (Richards, 2006).

1.3 Climate Change

Over the last three decades, the earth's average temperature has increased by about 0.7 °C. In the last 12 years, record high average temperatures were recorded. In Europe, 35,000 lives and tremendous amounts of grain harvest were lost due to extreme temperatures in August 2003 (Fischer et al., 2007). If this trend continues, in this century we will witness earth's temperature increase by 1.4–5.8 °C. However, the land and ocean temperature increase will be much greater than the average earth's temperature (Tesler and Langridge, 2010).

1.4 High-Input Agriculture

Use of chemical fertilizers has been a critical factor in the increased efficiency of cultivable land area for food production. In addition to increased fertilizer utilization, mechanization of agricultural operations like tilling and harvesting has also significantly improved food production, storage, and distribution. These mechanization practices, however, are raw material and infrastructure dependent (Tscharntke et al., 2012).

The most important concern in high-input agriculture is the loss of topsoil. An average of about 20 metric tons of topsoil is lost per hectare per year in the United States. This can further lead to loss of organic matter, water, and rooting penetration. The extensive nitrate and pesticide use leaches into the groundwater, which may cause health hazards (van Bruggen, 1995). Intensive mono-cropping systems favor epidemics caused by fungi, bacteria, nematodes, and other microorganisms. As a single crop is grown for extended periods on a certain soil, deleterious rhizobacteria may develop. Using an intra- or interspecific crop mixture as well as the use of barriers between crops can reduce the spread of some diseases.

Intensive agriculture, over extended periods of time, may also reduce the carbon content in the soil. The carbon chemistry of the soil can also change. It is possible to reverse this change in soil carbon levels and chemistry, but the extent and quality of this recovery are still not known (Cusack et al., 2013). Around 10 billion U.S. dollars are spent annually on pesticides used for crop protection (Oerke, 2006). Despite the fact that synthetic pesticides provide efficient control over several pests, there is great concern of building up resistance/tolerance to these chemical compounds due to continuous and indiscriminate pesticide use. Additionally, various unfavorable and damaging consequences for the environment, such as pollution of soil and water, may occur from overuse of these chemicals (Sharma et al., 2008; Syan et al., 2014). Most pesticides are not meant to be specific to a particular organism. Thus, many nontarget organisms, which may be useful to the crop, are also exterminated during crop protection implementations. Moreover, residual pesticide in raw food products has shown to cause health risks in humans (Khera et al., 2013).

2. Success Stories: The Green Revolution

In an investigation of global crop production by the Food and Agriculture Organization of the United Nations, it was seen that between 1965 and 1985, crop yields increased by 56%, compared to an increase of only 28% from 1985 to 2005. This was termed the Green Revolution, where global crop production was significantly increased by a combination of better genetics of plants and improved management practices such as irrigation, fertilizers, soil conservation, pesticide use, and farm mechanization.

The Green Revolution was successful because F1 hybrid varieties of maize and semi-dwarf, disease-resistant varieties of wheat and rice were developed using conventional breeding. These varieties were tolerant of intensive irrigation and fertilizer use without the risk of lodging or having rust epidemics (Evenson and Gollin, 2003). The varieties were also improved to be adaptive to different environmental conditions. The traditional rice cultivars were photoperiod sensitive and had been selected for specific growing conditions; thus their growth was restricted to just one season. The new improved varieties were made insensitive to photoperiod by introduction of an insensitivity gene and were able to grow in most seasons of subtropical countries. Moreover, the availability of genotypes that had a short growing duration was able to tremendously increase food supplies in several countries. By incorporating early-flowering genes for quicker maturity, the growing periods were reduced from 160–170 days to 110–115 days. Genes that conferred resistance to biotic stresses as well as tolerance to abiotic stresses were successfully incorporated in rice, wheat, and maize varieties (Khush, 2001).

The traditional method of overcoming stresses was to select plants with favorable alleles, which would confer stress tolerance in a plant. Wild germplasms, which contain resistance genes, were introduced in cultivated crops (Khush, 2001). In India, the Green Revolution was successful in the sense that although food deprivation and malnutrition continue to be endemic problems in rural areas, India has been able to attain autonomy in food. India is no longer reliant on imports or assistance from various other countries to feed its people. At the same time, expansion of the nonfood sector of the agricultural segment benefits the development of textiles, food processing, and various other agriculture-based industries. The extent to which Indian agriculture was transformed during the first three decades succeeding the Green Revolution is rather extraordinary. However, in the current agricultural phase, the influence of previous developments of the Green Revolution are not only fading but also posing new challenges to the environment; therefore, to sustain the ever growing population and environmental protection, another high-tech revolution would be absolutely necessary (Conway and Barbier, 1990; Ray and Ghose, 2014).

In the Green Revolution era, conventional plant breeding provided excellent resources for the development of new varieties and novel germplasm. Regardless of the value of the past contributions, classical breeding alone will not provide adequate breakthroughs to increase yield and quality, to solve the complex problems of biotic and abiotic stresses, and to develop/enhance bioeneregy capabilities of plant species. In the modern era, plant improvement seems vulnerable and inadequate without the contributions of the new tools of molecular biology and genomics. The new tools of genomics have great potential to unravel the complexity of genomes with respect to gene function.

3. Current Scenario: The Gene-Sequencing Revolution

Genomics provides innovative, integrative approaches to plant biochemistry, development, and physiology irrespective of species reproduction barriers. Genomics can play an important role to match the demand for food security in many ways, such as by increasing crop yield, conferring disease resistance, and increasing the nutritive value of food (Varshney et al., 2011). Whole-genome sequencing is a particularly important step toward crop improvement. The next-generation sequencing platform is the biggest advancement in technology in recent years. This technology made the sequencing of a whole genome or whole transcriptome more accessible with reduced cost and time. Advances in genomic technologies led to the full-genome sequencing of more than 100 plant species. Availability of sequence information made gene assessment an easy task in the case of polyploid crops with complex genomes such as hexaploid wheat.

3.1 Next-Generation Sequencing

Next-generation sequencing (NGS) is the biggest advancement of technology that evolved during recent years; it brought about a revolution in the field of plant genetics and biotechnology. NGS greatly reduces the sequencing time, cost, and labor as compared to other sequencing approaches. This strategy has led to the sequencing of 40 plant species since 2010. Extensive databases are publically available that provide instant help to find gene structure and function (Dolezel et al., 2014). NGS techniques include Illumina HiSeq/MiSeq, Life technologies Ion Torrent/Ion proton, Life Technologies SOLiD, and Roche 454. NGS is based on reversible sequencing chemistry and produces millions of short reads with a size of 50–250 bp (Abel and Duncavage, 2013). NGS technology also revolutionized the understanding of genome-wide genetic diversity by sequencing hundreds of accessions of cereal crops (Edwards et al., 2013). Advancements in NGS contributed to generate extensive, high-throughput, genome-wide molecular markers such as simple sequence repeats (SSRs), single nucleotide polymorphisms (SNPs), and copy number variations (CNVs). Discovery of cost-effective high-throughput molecular makers assisted the identification of genes/quantitative trait loci (QTLs) controlling complex traits. NGS tools are a boon for the identification of genetic variations such as translocations and inversions at the whole-genome level (Gupta and Varshney, 2013).

The availability of low-cost NGS platforms encouraged the whole-genome sequencing of less studied crops such as pigeon peas, which were sequenced using the Illumia platform. An abundance of genome-wide marker data through NGS led to the innovation of new breeding methods such as marker-assisted recurrent selection (MARS), genomic selection (GS), and genome-wide selection (GWS) to phenotype prediction from genotypic data in the case of complex traits controlled by QTLs (Varshney et al., 2013). Recently, NGS technology has been used to understand the domestication-related changes in the genomes of different rice cultivars. The analysis resulted in the identification of millions of SNPs and thousands of INDELs (insertions and deletions) between three rice cultivars (Sakai et al., 2014). NGS has potential applications in resequencing of plant genomes to identify genomic variations for association studies of complex traits. NGS promoted the availability of whole-genome sequences and public databases such as Phytozome (http://www.phytozome.net/), which provides access to 41 annotated and sequenced plant genomes. The latest release of "Ensemble plants" provides sequences and annotation of 26 plant species including wheat and barley (http://plants.ensembl.org/index.html) (Mochida and Shinozaki, 2013). NGS can also be an important tool in determining the methylation state of genomes to study the epigenetic control of genes (Mittler and Shulaev, 2013).

3.2 Comparative Genomics

Comparative genomics is based on collinearity and synteny of genes or chromosomes in diverse species descended from a common ancestor (Poursarebani et al., 2013). Comparative genomics studies provide us with the information about orthologous gene functions from different species that are expected to produce similar phenotypes. With the progress of sequencing facilities and the availability of whole-genome sequences for major cereals such as rice, maize, and barley, it is now possible to identify genes and predict their functions in those cereal crops in which their sequencing information is still limited. Comparative genomics predicts the gene function by exploring genomics and postgenomic associations for the genes within plant species or between plants and prokaryotes. The transcriptomics and proteomics data provide important postgenomic evidences of similarity; thus coexpression data from microarray or ribonucleic acid (RNA)-seq can be utilized for prediction of gene function. Biochemical functions can also be determined using 3D structures (Bradbury et al., 2013).

Availability of large-scale genomic information and conserved synteny between various grass species provides an opportunity to explore the gene function and structure (Mochida and Shinozaki, 2013). Comparative genomics has also emerged as an important tool for the identification of micro-RNA (mi-RNA) targets that are conserved during evolution and expected to play essential roles (Friedman and Burge, 2014). Sequence comparison using online resources such as "gramene" (http://www.gramene.org/) is an important comparative functional genomics analysis tool for crop plants (Monaco et al., 2014). Comparative analysis of RNA-seq expression profiling of watermelon resulted in the identification of genes homologous to tomato controlling carotenoid synthesis (Grassi et al., 2013). Comparative analysis of *Arabidopsis*, rice, barley and maize genomes permitted identification of several important gene families including *Sm* and *WAK* (Zhang et al., 2005; Kaur et al., 2013; Chen and Cao, 2014).

3.3 Functional Genomics

Functional genomics is a wide approach for predicting functions and interactions of genes and their products. As described in the previous section, the advancement of genome-sequencing platforms has made it possible to fully sequence a large number of plant genomes. An explosion of gene sequence information has posed a major challenge of identifying genes and determining their function. The genomics era now took a vital turn and entered the arena of functional genomics to address several fundamental questions about the role and working of genes at different levels with tissues specificity. However, integration and analysis of the genomic data is the biggest challenge these days (Mittler and Shulaev, 2013). There are a number of online web servers to annotate the gene sequencing using bioinformatics pipelines (Lohse et al., 2014). With sequencing, high-throughput SNP arrays can be designed to characterize QTLs in genome-wide association studies (Chen et al., 2014a). However, functional annotation of genes is an important exercise for gene regulation and its modification (Mieulet et al., 2013). Several reverse genetics tools, such as transposons mutagenesis, T-DNA (transfer deoxyribonucleic acid) insertion, RNA interference (RNAi), and targeting-induced local lesions in genomes (TILLING), enable researchers to study specific genes and their functions (Chen et al., 2014b). Introduction of the maize *Ac-Ds* transposable element system as a transposon tagging tool into heterologous species offers unprecedented opportunities to link genes with function by creating and characterizing mutant alleles (Singh et al., 2006, 2012). Similarly, virus-induced gene silencing has been considered as a rapid and cost-effective functional analysis tool for complex crop species (Stratmann and Hind, 2011). TILLING (targeting-induced local lesions in genomes) is another widely accepted reverse genetic approach that is currently being used to screen population for mutations in target genes (Lee et al., 2014).

Alternatively, serial analysis of gene expression (SAGE), massively parallel signature sequencing (MPSS), and micro- and macroarrays, are available in crop plants for simultaneous prediction of messenger RNA (mRNA) profile to track the activity of a large number of genes (Gupta et al., 2008). In addition to the above-mentioned molecular strategies, biochemical tools such as proteomics and metabolomics are also playing an important role to track the gene function through protein profile and metabolite expression, respectively (Gupta et al., 2010).

Rice 2020, a project launched by the International Rice Functional Genomics Steering Committee (IRFGSC), was aimed to coordinate the functional genomics research in field of rice. The main focus was to identify the functions of every gene in the rice genome for increasing the rice production to meet the challenge of food security (Zhang et al., 2008). Prediction of gene function through functional genomics will increase the food production and nutritional quality through the introduction of important genes in new varieties (Khush, 2005). Utilization of such approaches assisted in the identification of key genes controlling preharvest sprouting, an agronomic trait with economic value in cereals (Singh and Singh, 2012; Singh et al., 2013). With the continuous progress in functional genomics tools, breeders can develop the new varieties tolerant to various biotic and abiotic stresses (Kim et al., 2014). Recently, two varieties of Ug99 stem rust resistant wheat were developed using TILLING (Till et al., 2014). Availability of these recent genomics integrated platforms will make the development of new varieties reliable and efficient.

4. Molecular and Genomic-Assisted Breeding

Increasing the frequency of favorable alleles and selection of better-adopted crop species is a major goal of plant breeding. The availability of molecular and genomics tools has increased the efficiency of breeding during recent years. Molecular and genomics-assisted breeding technologies such as marker-assisted backcrossing (MABC) and marker-assisted recurrent selection (MARS) greatly facilitated selection of complex but important traits controlled by major genes or QTLs for the development of improved cultivars.

4.1 Marker-Assisted Backcrossing

Advanced time- and cost-effective breeding technologies are required to overcome the major constraints of food security (Xu et al., 2013). Molecular breeding, brought about through the use of molecular markers, is advancement over the traditional breeding methods. This leads to the selection of cultivars with improved traits using suitable markers (Panigrahi, 2013). Marker-assisted backcrossing (MABC) is the process of insertion favorable alleles from one genetic background to other through the use of molecular markers (Gupta and Varshney, 2013). There are several factors that affect the efficiency of MABC, such as selection strategy, population size, and marker density (Prigge et al., 2009). MABC has cut the breeding cycles in half by recovering the recurrent parent genome up to 98.7% after three backcrosses. Molecular markers tightly linked to the trait of interest are essential to perform MABC (Soto-Cerda et al., 2013). Elite chickpea cultivars resistant to *Fusarium* wilt (FW) and *Ascochyta* blight (AB) were developed through MABC using three backcrosses and three rounds of selfing (Varshney et al., 2014).

A simplified MABC technique with only one generation of backcross and phenotypic evaluation was used to improve hybrid rice for bacterial blight resistance (Ji et al., 2014). A major QTL for submergence tolerance was incorporated into a rain-fed lowland variety through two backcrosses and one selfing generation (Iftekharuddaula et al., 2010). Marker-assisted foreground selection and background selection is an important strategy for the traits controlled by major genes. However, this strategy has been found to be inefficient for QTLs with small effects and also for the breeding populations under different environmental conditions (Xu et al., 2013).

4.2 Marker-Assisted Recurrent Selection

Yield increase and biotic/abiotic stress resistance are important goals toward food security that can be achieved through new breeding practices. Marker-assisted recurrent selection (MARS) is an important strategy for the traits controlled by a moderately large number of QTLs (Xu et al., 2013). MARS refers to the improvement of an F2 population by one cycle of phenotypic and marker data followed by two or three cycles of only marker data. It is a process of pyramiding favorable alleles from different genetic sources (Gupta and Varshney, 2013). MARS can also be performed without prior knowledge of QTLs (Xu et al., 2012). Desirable alleles at several target loci were incorporated through MARS in crops such as sunflowers, soybeans, and maize (Tester and Langridge, 2010). Diverse molecular markers are a prerequisite for molecular breeding techniques. There are different classes of molecular markers based on the level of functional characterization.

4.3 Random Markers

Random markers (RDMs) are derived arbitrarily from polymorphic sites in genomic DNA (deoxyribonucleic acid) and complementary DNA (cDNA) (Gupta and Rustgi, 2004). RDM markers are mainly developed using restriction enzyme-based methods. The most commonly used random DNA markers are restriction fragment length polymorphisms (RFLPs), simple sequence repeats (SSRs), and amplified fragment length polymorphisms (AFLPs). Sequence information is required for SSRs and SNPs but not for RFLPs, random amplified polymorphic DNA (RAPD), AFLPs, etc. SSRs and SNPs are markers of choice for molecular breeding (Salgotra et al., 2014). These makers play an important role in crop improvement in many ways, such as gene introgression via MABC/MARS, germplasm characterization, diversity analysis, identifying polymorphisms, construction of molecular maps, QTL analysis, gene tagging, map-based cloning, and phylogenetic analysis (Varshney et al., 2007).

4.4 Gene Target Markers

Due to the availability of high-throughput sequencing platforms and genomic information, the trend of markers shifted from RDMs to gene target markers (GTMs) and functional markers (FMs) located in or near the gene of interest (Poczai et al., 2013). GTMs are developed from polymorphic sites within genes that may or may not be involved in phenotypic trait variations (Varshney et al., 2007). These markers can also tag untranslated regions of genes (Poczai et al., 2013). These markers have been developed through sequencing, expression profiling, sequence comparisons, or synteny studies (Andersen and Lubberstedt, 2003).

4.5 Functional Markers

Genome-wide sequencing provides the platform to mine molecular markers (Muthamilarasan et al., 2013). Functional markers (FMs) are the polymorphic sites within genes that are functionally validated for phenotypic variations (Salgotra et al., 2014). These markers are further classified into two groups: indirect functional markers (IFMs) and direct functional markers (DFMs) depending upon indirect or direct proof of their role in phenotypic trait variation. Functional markers can be derived from nonredundant EST database either by direct mapping or database mining. GTMs and FMs permit detection of the nucleotide diversity in the genes that control agronomic traits. These markers are useful to predict the genetic relationship as well as functional diversity in relation to adaptive variation. In contrast to RDMs, these markers are transferable to related species or genera (Varshney et al., 2007).

5. Genomics-Integrated Breeding Technologies

Food security is the major concern in an exponentially growing world population. There is need to develop high-yielding crops with durable tolerance to biotic and abiotic stresses. The modern era of -omics (functional genomics, comparative genomics) and high-throughput marker technology provides an opportunity to understand the functions of genes with small effects that underlie most of the important traits. Genome-wide markers have the potential to capture all additive effects for selection of desirable genotypes. Emerging genomic-integrated breeding technologies are revolutionizing the understanding about mechanisms of complex quantitative yield and stress tolerance traits in a time/cost-efficient manner.

5.1 Genome-Wide Association

Genetic improvement of crop productivity is one of the many solutions to increase food security (Brachi et al., 2011). Around 50% of the increase in grain yield has been attributed to genetic improvement. Genome-wide association (GWA) based on large-scale sequencing provides a powerful tool to find genetic variants that can increase crop yield. It is an important strategy to identify genes underlying variations in quantitatively inherited traits. These genes may work in an orchestrated manner and increase grain yield under various stress conditions (Xue et al., 2013).

There are number of factors influencing the association mapping studies, such as genetic marker coverage, number of individuals studied, and linkage disequilibrium (Cockram et al., 2010). Marker density on a genomic map should be higher than the extent of linkage disequilibrium (LD), which in turn depends upon the population structure, genetic diversity, and number of recombination events that happened to restructure that diversity (Brachi et al., 2011). This approach is based on the principle of LD. Whole-genome scanning for LD between mapped marker loci and traits of interest is called GWA (Neumann et al., 2010).

A study integrating the approach of sequence-based GWA and functional genome annotation displayed the potential of matching complex traits to their causal polymorphisms in rice (Huang et al., 2012). Modern maize breeding techniques have shown a remarkable increase in its productivity in the last few decades. As maize is such a diverse crop, a genome-wide genetic variation pattern among various maize lines has been studied extensively. In a GWA maize study, two candidate genes were identified that were associated with yield-related traits measured under water-stress conditions (Hu and Xiong, 2013). A nested association mapping (NAM) population of 25 RIL families was generated for quantitative trait analysis in maize (McMullen et al., 2009). A GWA study of the maize NAM panel was performed to determine the genetic basis of quantitative leaf architecture traits and identification of some of the important genes (Tian et al., 2011). GWA studies found a strong association between genetic loci and 14 agronomic traits in the population of *Oryza sativa* subspecies *indica* (Huang et al., 2010). Genetic architecture of (aluminum) Al tolerance and Al tolerance loci in rice was identified through biparental QTL mapping and GWA (Famoso et al., 2011).

5.2 Genomic Selection

Traditional plant breeding programs are primarily based on phenotypic evaluation and pedigree information. This strategy is not able to predict the true genetic potential of an individual where environmental factors are masking the genetic effects. Moreover, the pedigree-based selection methods narrowed the genetic diversity through the selection of more related individuals (Lin et al., 2014). Marker-assisted selection (MAS) uses the molecular markers for the selection of a trait of interest. Marker-assisted foreground selection and background selection is an important strategy for the traits controlled by major genes.

Although MAS can predict the breeding values more accurately than traditional methods, this strategy only accounts for major genes or QTLs, leaving behind the important genes with minor effects and genomic regions involved in complex traits. This strategy is not very efficient for the breeding populations under different environmental conditions. Linkage drag (linkage of undesirable genes with the gene controlling the trait of interest) is another major problem in the introgression of genes into elite cultivars through MAS. Genetic background effects and allelic diversity in the breeding programs cannot be captured through a biparental population (Collard and Mackill, 2008; Heffner et al., 2009). Moreover, multiple cycles of backcrossing are required to recover the maximum recurrent parent genotype through MAS (Bohra, 2013).

On the other hand, genomic selection (GS) based on LD can be applied to the populations having extensive phenotypic data over the years to dissect complex traits. This process also avoids the generation of special mapping populations (Xu et al., 2013). In contrast to a few major genes/QTLs, thousands of molecular markers possessing strong LD with the trait of interest are used for GS. A number of simulation studies in various crops, such as maize, oil palm (Bernardo and Yu, 2007; Wong and Bernardo, 2008), and forages (Simeao Resende et al., 2014), illustrated higher genetic gain through GS as compared to MAS or phenotypic selection. GS predicts the breeding values based on phenotyping and genotyping of only a small training population and selection is based on the genotyping of a breeding population at early stages without phenotyping.

Genomic selection was first introduced by (Meuwissen et al., 2001) as an advancement in molecular breeding technology for the study of quantitative traits. Quantitative traits are controlled by a large number of genes, with a cumulative effect of each gene on the trait. This approach uses whole-genome molecular markers (high-density markers and high-throughput genotyping) to develop a prediction model for estimating a breeding value for each individual (Crossa et al., 2011). The availability of full-genome sequences through NGS (next generation sequencing) has provided high-throughput molecular markers (Jonas and de Koning, 2013).

The prediction model attempts to capture the total additive genetic variance, in contrast to MARS, which uses only a few markers (Rutkoski et al., 2013). A training population (TP) is genotyped with a large number of markers and phenotyped for important traits. In a breeding population, individuals are genotyped but not phenotyped. A genomic estimated breeding value (GEBV) for each individual is obtained by summing marker effects for that genotype (Heffner et al., 2009). This approach leads to the estimation of marker/QTL effects for individuals in a breeding population without phenotyping. In this method, the markers should be in LD with the target QTLs affecting the trait. The minimum number of markers required for genotyping depends on the LD decay. The rate of LD decay is determined by several factors, such as population size, recombination rates, evolutionary history of the population, and selection efforts (Charmet and Storlie, 2012).

This approach is far better than MABC/MARS for improving complex traits with low heritability. GS is able to solve the problems related to QTL number, such as allele distribution and epistatic effects. GS is three times more efficient than phenotypic selection in terms of processing time. It reduces the breeding cycle time from 9 years to 3 years by eliminating the need for repeated phenotypic selections. In wheat breeding programs, the GS with GEBV accuracies of only 0.5 could lead to a twofold higher gain per year as compared to MAS (Desta and Ortiz, 2014).

Simulation studies estimated a correlation of 0.85 between true breeding values and GEBV for traits controlled by multiple genes and having low heritability (Heffner et al., 2009). Breeding values of a wheat population were predicted with high accuracy using diversity arrays technology (DArT) markers to complement phenotypic selection (Charmet and Storlie, 2012). The accuracy of GEBV for rust resistance in wheat was validated using 206 diverse wheat landraces.

Moderate prediction accuracies were achieved through genomic selection (Daetwyler et al., 2014). GS studies that used empirical data from maize, barley, and wheat have been performed to test the accuracy of GEBV predictions for different traits (Nakaya and Isobe, 2012). The GS models were developed using a TP consisting of International Maize and Wheat Improvement Center (CIMMYT) materials with varying levels of stem rust adult plant resistance.

6. Implementation of Integrated Breeding Strategies for Food Security

Despite the fact that genomics-assisted breeding has been successfully implemented and is available in several crops, plant breeders are not routinely using these strategies. This is especially true in the public breeding sector (Ribaut et al., 2010). The reason for such lack in integration of these strategies with conventional approaches could be due to unsuitable phenotyping setup, lack of trained next-generation molecular personnel, and inadequate bioinformatics knowledge (Tester and Langridge, 2010).

Genomics-assisted breeding approaches are effective in developing productive crops with resistance to biotic stresses in a much more time- and cost-effective way when compared to their conventional breeding counterparts (Varshney et al., 2009). In the future, it is expected that historic crop phenotypic data will be publicly available. This would permit plant scientists to share their results and to compare their phenotypes to those deposited previously and be able to identify new, and occasionally rare, alleles that could improve crop productivity (Figure 1). Public sharing of genomic and

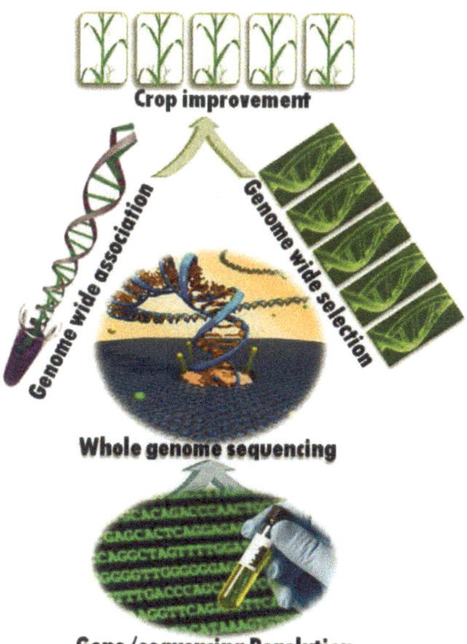

Figure 1 Sequencing-based crop improvement strategy.

phenotypic data will return faster results in terms of increasing crop yield. Eventually, integrated breeding including bioinformatics, molecular approaches, and conventional methods will have to be followed by researchers toward crop improvement for stress tolerance to conditions such as salinity, drought, and fluctuating environmental conditions. These integrated methods will ensure food security as well as sustainable agriculture, primarily in developing countries.

References

Abel, H.J., Duncavage, E.J., 2013. Detection of structural DNA variation from next generation sequencing data: a review of informatic approaches. Cancer Genetics 206 (12), 432–440.

Andersen, J.R., Lubberstedt, T., 2003. Functional markers in plants. Trends in Plant Science 8 (11), 554–560.

Bernardo, R., Yu, J., 2007. Prospects for genomewide selection for quantitative traits in maize. Crop Science 47 (3), 1082–1090.

Bohra, A., 2013. Emerging paradigms in genomics-based crop improvement. The Scientific World Journal 2013, 585467.

Brachi, B., Morris, G.P., Borevitz, J.O., 2011. Genome-wide association studies in plants: the missing heritability is in the field. Genome Biology 12 (10), 232.

Bradbury, L.M., Niehaus, T.D., Hanson, A.D., 2013. Comparative genomics approaches to understanding and manipulating plant metabolism. Current Opinion in Biotechnology 24 (2), 278–284.

Brown, L.R., 2009. Could food shortages bring down civilization? Scientific American 300, 50–57.

Chang, C.C., Lee, H.L., Hsu, S.H., 2013. Food security: global trends and regional perspective with reference to East Asia. Pacific Review 5, 589–613.

Charmet, G., Storlie, E., 2012. Implementation of genome-wide selection in wheat. Russian Journal of Genetics: Applied Research 2 (4), 298–303.

Chen, H., Xie, W., He, H., Yu, H., Chen, W., Li, J., Yu, R., Yao, Y., Zhang, W., He, Y., et al., 2014a. A high-density SNP genotyping array for rice biology and molecular breeding. Molecular Plant 7 (3), 541–553.

Chen, L., Hao, L., Parry, M.A., Phillips, A.L., Hu, Y.G., 2014b. Progress in TILLING as a tool for functional genomics and improvement of crops. Journal of Integrative Plant Biology 56 (5), 425–443.

Chen, Y., Cao, J., 2014. Comparative genomic analysis of the Sm gene family in rice and maize. Gene 539 (2), 238–249.

Cockram, J., White, J., Zuluaga, D.L., Smith, D., Comadran, J., Macaulay, M., Luo, Z., Kearsey, M.J., Werner, P., Harrap, D., et al., 2010. Genome-wide association mapping to candidate polymorphism resolution in the unsequenced barley genome. Proceedings of the National Academy of Sciences of the United States of America 107 (50), 21611–21616.

Collard, B.C., Mackill, D.J., 2008. Marker-assisted selection: an approach for precision plant breeding in the twenty-first century. Philosophical Transactions of the Royal Society of London Series B Biological Sciences 363 (1491), 557–572.

Conway, G.R., Barbier, E.B., 1990. After the Green. Revolution Earthscan, London.

Crossa, J., Pérez, P., de los Campos, G., Mahuku, G., Dreisigacker, S., Magorokosho, C., 2011. Genomic selection and prediction in plant breeding. Journal of Crop Improvement 25 (3), 239–261.

Cusack, D.F., Chadwick, O.A., Ladefoged, T., Vitousek, P.M., 2013. Long-term effects of agriculture on soil carbon pools and carbon chemistry along a Hawaiian environmental gradient. Biogeochemistry 112, 229–243.
Daetwyler, H.D., Bansal, U.K., Bariana, H.S., Hayden, M.J., Hayes, B.J., 2014. Genomic prediction for rust resistance in diverse wheat landraces. TAG. Theoretical and Applied Genetics. Theoretische und Angewandte Genetik 127 (8), 1795–1803.
Desta, Z.A., Ortiz, R., 2014. Genomic selection: genome-wide prediction in plant improvement. Trends in Plant Science 19 (9), 592–601.
DeVilliers, S.M., Hoisington, D.A., 2011. The trends and future of biotechnology crops for insect pest control. African Journal of Biotechnology 10, 4677–4681.
Dhaliwal, G.S., Jindal, V., Dhawan, A.K., 2010. Insect pest problems and crop losses: changing trends. Indian Journal of Ecology 37 (1), 1–7.
Dolezel, J., Vrana, J., Capal, P., Kubalakova, M., Buresova, V., Simkova, H., 2014. Advances in plant chromosome genomics. Biotechnology Advances 32 (1), 122–136.
Edwards, D., Batley, J., Snowdon, R.J., 2013. Accessing complex crop genomes with next-generation sequencing. TAG. Theoretical and Applied Genetics. Theoretische und Angewandte Genetik 126 (1), 1–11.
Evenson, R.E., Gollin, D., 2003. Assessing the impact of the green revolution, 1960–2000. Science 300, 758–762.
Famoso, A.N., Zhao, K., Clark, R.T., Tung, C.W., Wright, M.H., Bustamante, C., Kochian, L.V., McCouch, S.R., 2011. Genetic architecture of aluminum tolerance in rice (*Oryza sativa*) determined through genome-wide association analysis and QTL mapping. PLoS Genetics 7 (8), e1002221.
Fischer, G., Tubiello, F.N., van Velthuizen, H., Wiberg, D., 2007. Climate change impacts on irrigation water requirements: effects of mitigation, 1990–2080. Technological Forecasting and Social Change. 74, 1083–1107.
Foley, J.A., et al., 2011. Solutions for a cultivated planet. Nature 478, 337–342.
Friedman, R.C., Burge, C.B., 2014. MicroRNA target finding by comparative genomics. Methods in Molecular Biology (Clifton, NJ) 1097, 457–476.
Godfray, H.C.J., et al., 2010. Food security: the challenge of feeding 9 billion people. Science 327, 812–818.
Grassi, S., Piro, G., Lee, J.M., Zheng, Y., Fei, Z., Dalessandro, G., Giovannoni, J.J., Lenucci, M.S., 2013. Comparative genomics reveals candidate carotenoid pathway regulators of ripening watermelon fruit. Bmc Genomics 14.
Gupta, P., Langridge, P., Mir, R., 2010. Marker-assisted wheat breeding: present status and future possibilities. Molecular Breeding 26 (2), 145–161.
Gupta, P., Mir, R., Mohan, A., Kumar, J., 2008. Wheat genomics: present status and future prospects. International Journal of Plant Genomics. 2008. 10.1155/2008/896451.
Gupta, P., Rustgi, S., 2004. Molecular markers from the transcribed/expressed region of the genome in higher plants. Functional & Integrative Genomics 4 (3), 139–162.
Gupta, P.K., Varshney, R.K., 2013. Cereal Genomics II. Springer.
Heffner, E.L., Sorrells, M.E., Jannink, J.-L., 2009. Genomic selection for crop improvement. Crop Science 49 (1), 1–12.
Hu, H., Xiong, L., 2013. Genetic engineering and breeding of drought-resistant crops. Annual Review of Plant Biology 65, 715–774.
Huang, X., Wei, X., Sang, T., Zhao, Q., Feng, Q., Zhao, Y., Li, C., Zhu, C., Lu, T., Zhang, Z., 2010. Genome-wide association studies of 14 agronomic traits in rice landraces. Nature Genetics 42 (11), 961–967.
Huang, X., et al., 2012. Genome-wide association study of flowering time and grain yield traits in a worldwide collection of rice germplasm. Nature Genetics 44, 32–39.

Iftekharuddaula, K.M., Newaz, M.A., Salam, M.A., Ahmed, H.U., Mahbub, M.A.A., Septiningsih, E.M., Collard, B.C.Y., Sanchez, D.L., Pamplona, A.M., Mackill, D.J., 2010. Rapid and high-precision marker assisted backcrossing to introgress the SUB1 QTL into BR11, the rainfed lowland rice mega variety of Bangladesh. Euphytica 178 (1), 83–97.

Ji, Z., Shi, J., Zeng, Y., Qian, Q., Yang, C., 2014. Application of a simplified marker-assisted backcross technique for hybrid breeding in rice. Biologia 69 (4), 463–468.

Jonas, E., de Koning, D.J., 2013. Does genomic selection have a future in plant breeding? Trends in Biotechnology 31 (9), 497–504.

Jones, A., 2001. Eating Oil: Food Supply in a Changing Climate. Sustain/Elm Farm Research Centre, London.

Kaur, R., Singh, K., Singh, J., 2013. A root specific wall-associated kinase gene, *HvWAK1*, regulates root growth, and is highly divergent in barley and other cereals. Functional & Integrative Genomics 13, 167–177.

Khera, P., Pandey, M.K., Varshney, R.K., 2013. Pest and diseases: old and new threats – modern breeding tools to tailor new crop cultivars. Secheresse 24, 261–273.

Khush, G.S., 2001. Green revolution: the way forward. Nature Reviews Genetics 2, 815–822.

Khush, G.S., 2005. What it will take to feed 5.0 billion rice consumers in 2030. Plant Molecular Biology 59 (1), 1–6.

Kim, D.-W., Agrawal, G.K., Rakwal, R., Ahmed, S., Rohila, J.S., 2014. Genomic methods for improving abiotic stress tolerance in crops. In: Plant Biotechnology. Springer, pp. 35–42.

Lee, L.S., Till, B.J., Hill, H., Huynh, O.A., Jankowicz-Cieslak, J., 2014. Mutation and mutation screening. Methods in Molecular Biology (Clifton, NJ) 1099, 77–95.

Lin, Z., Hayes, B.J., Daetwyler, H.D., 2014. Genomic selection in crops, trees and forages: a review. Crop and Pasture Science 65 (11), 1177–1191.

Lohse, M., Nagel, A., Herter, T., May, P., Schroda, M., Zrenner, R., Tohge, T., Fernie, A.R., Stitt, M., Usadel, B., 2014. Mercator: a fast and simple web server for genome scale functional annotation of plant sequence data. Plant, Cell & Environment 37 (5), 1250–1258.

McMullen, M.D., Kresovich, S., Villeda, H.S., Bradbury, P., Li, H., Sun, Q., Flint-Garcia, S., Thornsberry, J., Acharya, C., Bottoms, C., et al., 2009. Genetic properties of the maize nested association mapping population. Science 325 (5941), 737–740.

Meuwissen, T.H.E., Hayes, B.J., Goddard, M.E., 2001. Prediction of total genetic value using genome-wide dense marker maps. Genetics 157 (4), 1819–1829.

Mieulet, D., Meynard, D., Sire, C., Petit, J., Breitler, J.-C., Perin, C., Dievart, A., Divol-Malgoire, F., Courtois, B., Verdeil, J.-L., et al., 2013. Gene transfer: a tool for the functional analysis of the rice genome. Cahiers Agricultures 22 (5), 484–493.

Mittler, R., Shulaev, V., 2013. Functional genomics, challenges and perspectives for the future. Physiologia Plantarum 148 (3), 317–321.

Mochida, K., Shinozaki, K., 2013. Unlocking Triticeae genomics to sustainably feed the future. Plant & Cell Physiology 54 (12), 1931–1950.

Monaco, M.K., Stein, J., Naithani, S., Wei, S., Dharmawardhana, P., Kumari, S., Amarasinghe, V., Youens-Clark, K., Thomason, J., Preece, J., et al., 2014. Gramene 2013: comparative plant genomics resources. Nucleic Acids Research 42 (D1), D1193–D1199.

Muthamilarasan, M., Theriappan, P., Prasad, M., 2013. Recent advances in crop genomics for ensuring food security. Current Science 105, 155–158.

Nakaya, A., Isobe, S.N., 2012. Will genomic selection be a practical method for plant breeding? Annals of Botany 110 (6), 1303–1316.

Neumann, K., Kobiljski, B., Denčić, S., Varshney, R.K., Börner, A., 2010. Genome-wide association mapping: a case study in bread wheat (*Triticum aestivum* L). Molecular Breeding 27 (1), 37–58.

Oerke, E.C., Dehne, H.W., 2004. Safeguarding production losses in major crops and the role of crop protection. Crop Protection 23, 275–285.
Oerke, E.C., 2006. Crop losses to pests. Journal of Agricultural Science 144, 31–43.
Panigrahi, J., 2013. Marker-assisted breeding for simple inherited traits conferring stress resistance in crop plants. Development 25, 27.
Poczai, P., Varga, I., Laos, M., Cseh, A., Bell, N., Valkonen, J.P., Hyvönen, J., 2013. Advances in plant gene-targeted and functional markers: a review. Plant Methods 9 (1), 6.
Poursarebani, N., Ariyadasa, R., Zhou, R., Schulte, D., Steuernagel, B., Martis, M.M., Graner, A., Schweizer, P., Scholz, U., Mayer, K., et al., 2013. Conserved synteny-based anchoring of the barley genome physical map. Functional & Integrative Genomics 13 (3), 339–350.
Prigge, V., Melchinger, A.E., Dhillon, B.S., Frisch, M., 2009. Efficiency gain of marker-assisted backcrossing by sequentially increasing marker densities over generations. TAG. Theoretical and Applied Genetics. Theoretische und Angewandte Genetik 119 (1), 23–32.
Ray, S.C., Ghose, A., 2014. Production efficiency in Indian agriculture: an assessment of the post green revolution years. Omega 44, 58–69.
Ribaut, J.-M., de Vicente, M.C., Delannay, X., 2010. Molecular breeding in developing countries. Current Opinion in Plant Biology 13 (2), 213–218.
Richards, R.A., 2006. Physiological traits used in the breeding of new cultivars for water-scarce environments. Agricultural Water Management 80, 197–211.
Runge, C.F., Senauer, B., 2007. How biofuels could starve the poor. Foreign Affairs 86 (3), 41–53.
Rutkoski, J.E., Poland, J., Jannink, J.-L., Sorrells, M.E., 2013. Imputation of unordered markers and the impact on genomic selection accuracy. G3: Genesl Genomesl Genetics 3 (3), 427–439.
Sakai, H., Kanamori, H., Arai-Kichise, Y., Shibata-Hatta, M., Ebana, K., Oono, Y., Kurita, K., Fujisawa, H., Katagiri, S., Mukai, Y., et al., 2014. Construction of pseudomolecule sequences of the aus rice cultivar Kasalath for comparative genomics of Asian cultivated rice. DNA Research : An International Journal for Rapid Publication of Reports on Genes and Genomes 21 (4), 397–405.
Salgotra, R.K., Gupta, B.B., Stewart Jr., C.N., 2014. From genomics to functional markers in the era of next-generation sequencing. Biotechnology Letters 36 (3), 417–426.
Schulman, A.H., 2012. Genomics Tools for Accelerating Plant Breeding. Maataloustieteen Päivät 2012, 10.-11.1. 2012 Viikki, Helsinki: esitelmät, posterit/Toim. Nina Schulman ja Heini Kauppinen.
Sharma, H.C., Varshney, R.K., Gaur, P.M., Gowda, C.L.L., 2008. Potential for using morphological, biochemical, and molecular markers for resistance to insect pests in grain legumes. Journal of Food Legumes 21, 211–217.
Simeao Resende, R.M., Casler, M.D., Vilela de Resende, M.D., 2014. Genomic selection in forage breeding: accuracy and methods. Crop Science 54 (1), 143–156.
Singh, J., Zhang, S., Chen, C., Cooper, L., Bregitzer, P., Sturbaum, A.K., Hayes, P.M., Lemaux, P.G., 2006. High-frequency Ds remobilization over multiple generations in barley facilitates gene tagging in large genome cereals. Plant Molecular Biology 62, 937–950.
Singh, M., Singh, J., 2012. Seed development related expression of ARGONAUTE 4_9 class of genes in barley: possible role in seed dormancy. Euphytica 188, 123–129.
Singh, M., Singh, S., Randhawa, H., Singh, J., 2013. Polymorphic homoeolog of key gene of RdDM pathway, ARGONAUTE4_9 class is associated with pre-harvest sprouting in wheat (*Triticum aestivum* L.). PLoS One. 8 (10), e77009.

Singh, S., Tan, H.Q., Singh, J., 2012. Mutagenesis of barley malting quality QTLs with Ds transposons. Functional & Integrative Genomics 12 (1), 131–141.
Syan, H.S., Prasher, S., Pageau, D., Singh, J., 2014. Dissipation of major herbicides applied in transgenic and non-transgenic canola production in Quebec. European Journal of Soil Biology 63, 21–27.
Soto-Cerda, B.J., Peñaloza, E.H., Montenegro, A.B., Rupayan, A.R., Gallardo, M.H., Salvo-Garrido, H., 2013. An efficient marker-assisted backcrossing strategy for enhancing barley (*Hordeum vulgare* L.) production under acidity and aluminium toxicity. Molecular Breeding 31 (4), 855–866.
Stratmann, J.W., Hind, S.R., 2011. Gene silencing goes viral and uncovers the private life of plants. Entomologia Experimentalis Et Applicata 140 (2), 91–102.
Tester, M., Langridge, P., 2010. Breeding technologies to increase crop production in a changing world. Science 327 (5967), 818–822.
Tian, F., Bradbury, P.J., Brown, P.J., Hung, H., Sun, Q., Flint-Garcia, S., Rocheford, T.R., McMullen, M.D., Holland, J.B., Buckler, E.S., 2011. Genome-wide association study of leaf architecture in the maize nested association mapping population. Nature Genetics 43 (2), 159–162.
Till, B.J., Amos, E.K., Kinyua, M.G., Bado, S., Jankowicz-Cieslak, J., Huynh, O.A., Henry, I., Comai, L., Lagoda, P., 2014. Mutation induction and reverse-genetics for functional genomics and breeding. In: Tagungsband der 64 Jahrestagung der Vereinigung der Pflanzenzüchter und Saatgutkaufleute Österreichs 25–26 November 2013. Raumberg-Gumpenstein. 5.
Tscharntke, T., Clough, Y., Wanger, T.C., Jackson, L., Motzke, I., Perfecto, I., Vandermeer, J., Whitbread, A., 2012. Global food security, biodiversity conservation and the future of agricultural intensification. Biological Conservation 151 (1), 53–59.
Varshney, R.K., Bansal, K.C., Aggarwal, P.K., Datta, S.K., Craufurd, P.Q., 2011. Agricultural biotechnology for crop improvement in a variable climate: hope or hype? Trends in Plant Science 16, 363–371.
Varshney, R.K., Close, T.J., Singh, N.K., Hoisington, D.A., Cook, D.R., 2009. Orphan legume crops enter the genomics era!. Current Opinion in Plant Biology 12, 202–210.
Varshney, R.K., Mahendar, T., Aggarwal, R.K., Börner, A., 2007. Genic molecular markers in plants: development and applications. In: Genomics-Assisted Crop Improvement. Springer, pp. 13–29.
Varshney, R.K., Mohan, S.M., Gaur, P.M., Chamarthi, S.K., Singh, V.K., Srinivasan, S., Swapna, N., Sharma, M., Pande, S., Varshney, R.K., et al., 2014. Marker-assisted backcrossing to introgress resistance to Fusarium wilt Race 1 and ascochyta Blight in C 214, an elite cultivar of chickpea. The Plant Genome 7 (1).
Varshney, R.K., Mohan, S.M., Gaur, P.M., Gangarao, N.V., Pandey, M.K., Bohra, A., Sawargaonkar, S.L., Chitikineni, A., Kimurto, P.K., Janila, P., et al., 2013. Achievements and prospects of genomics-assisted breeding in three legume crops of the semi-arid tropics. Biotechnology Advances 31 (8), 1120–1134.
van Bruggen, A.H.C., 1995. Plant disease severity in high-input compared to reduced-input and organic farming systems. Plant Disease 79, 976–984.
Wall, E., Smit, B., 2005. Climate change adaption in light of sustainable agriculture. Journal of Sustainable Agriculture 27, 113–123.
Wong, C., Bernardo, R., 2008. Genomewide selection in oil palm: increasing selection gain per unit time and cost with small populations. Theoretical and Applied Genetics 116 (6), 815–824.
Xu, Y., Lu, Y., Xie, C., Gao, S., Wan, J., Prasanna, B.M., 2012. Whole-genome strategies for marker-assisted plant breeding. Molecular Breeding 29 (4), 833–854.

Xu, Y., Xie, C., Wan, J., He, Z., Prasanna, B.M., 2013. Marker-Assisted Selection in Cereals: Platforms, Strategies and Examples. pp. 375–411.

Xue, Y., Warburton, M.L., Sawkins, M., Zhang, X., Setter, T., Xu, Y., Grudloyma, P., Gethi, J., Ribaut, J.-M., Li, W., Zhang, X., Zheng, Y., Yan, J., 2013. Genome-wide association analysis for nine agronomic traits in maize under well-watered and water-stressed conditions. Theoretical and Applied Genetics 126, 2587–2596.

Zhang, S., Chen, C., Li, L., Meng, M., Singh, J., Jiang, N., Deng, X.-H., He, Z.-H., Lemaux, P.G., 2005. Evolutionary expansion, gene structure, and expression of the rice (*Oryza sativa* L.) wall-associated kinase (*OsWAKs*) gene family. Plant Physiology 139, 1107–1124.

Zhang, Q., Li, J., Xue, Y., Han, B., Deng, X.W., 2008. Rice 2020: a call for an international coordinated effort in rice functional genomics. Molecular Plant 1 (5), 715–719.

Successful Technologies and Approaches Used to Develop and Manage Resistance against Crop Diseases and Pests

3

G.L. Hartman[1,2], M.L. Pawlowski[2], H.-X. Chang[2], C.B. Hill[3]
[1]USDA-Agricultural Research Service, University of Illinois, Urbana, IL, USA; [2]Department of Crop Sciences, University of Illinois, Urbana, IL, USA; [3]Agricen Sciences, LLC, Pilot Point, TX, USA

1. Introduction

Increases in crop production have been partially the result of increased resistance to pathogens and pests. There are many historical examples of how increased resistance to pathogens and pests have increased crop productivity in the major food crops like wheat (*Triticum aestivum* L.), maize (*Zea mays* L.), rice (*Oryza sativa* L.), and soybeans (Glycine max (L.) Merr.). Of these major food crops, wheat was harvested in more hectares worldwide than any other crop, followed by maize, rice, and soybeans in 2013 (Figure 1; www.faostat3.fao.org). A United Nations post from March 2014 indicated that production of 70% more food would be needed by 2050 to feed the projected 9.1 billion human population (www.populationinstitute.org; Alexandratos and Bruinsma, 2012). With the current technologies, food production has kept up with demands, but further challenges remain, and technologies that are emerging or yet to be proven or invented may be needed to produce enough food for demand in the next few decades and beyond.

1.1 World Losses Due to Pathogens and Pests

Increases in crop production due to genetic improvements are often offset by constraints in production caused by broadly categorized abiotic and/or biotic factors. Although this chapter focuses on biotic constraints, the challenges growers face in crop production is a total package that includes abiotic and biotic factors, and also other factors from land and seed availability to unpredictable weather patterns to fertilizer availability and variable soil quality. Biotic constraints, such as pathogens, arthropods, and weeds, are detrimental to production and result in significant negative impacts to yield. Broadly defined, biotic constraints were estimated to cause potential global losses of up to 29% in wheat, 31% for maize, 37% for rice, and 29% for soybeans (Oerke, 2005). Predicting losses due to pathogens and pests of crop plants is more of an art than a science, as there are many contributing variables, including severity of attack, environmental conditions, host plant susceptibility, stage of plant

Emerging Technologies for Promoting Food Security. http://dx.doi.org/10.1016/B978-1-78242-335-5.00003-2

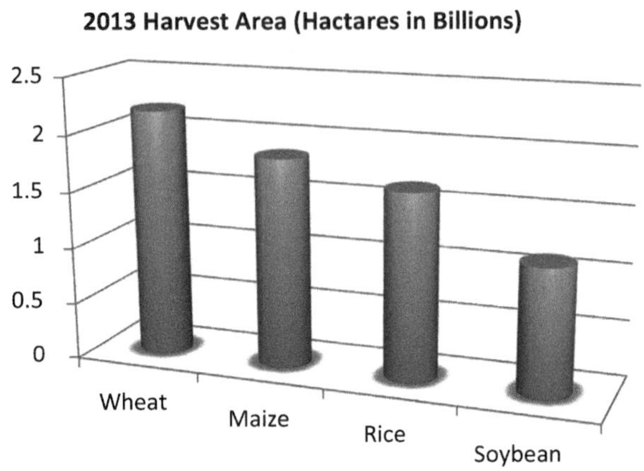

Figure 1 Production area of major crop plants.

development, and area affected, which are often not accurately measured, primarily due to insufficient technologies or high cost of conducting surveys. Regardless, much effort and expense has been directed toward protecting crop production from biotic stresses, especially in developed countries. There are a number of practices used alone or in combination to manage pathogens and pests of crop plants, including cultural techniques, pesticide applications, and deployment of resistance.

Along with increasing productivity of crop plants made through advances in breeding, there have been concerted efforts and advances made to increase resistance to pathogens and pests. This has primarily been based on the use and deployment of major resistance genes singly or by combining multiple resistance genes in gene pyramids, which broaden the spectrum of activity and increase durability. A more recent approach has been the development of transgenic plants that express exogenous genes controlling resistance. In addition, biological or chemical products that induce host resistance have recently been deployed into crop production as seed treatments, soil drenches, or topically applied sprays to help crop plants stand guard or respond to pathogen and pest attack.

In this introduction, a few examples are provided to show the significance of diseases and pests on the most widely harvested food crops and how the role of resistance has had a major impact on sustaining crop production. The sections following this introduction include approaches to increase resistance durability, improvement of crop resistance with genetic transformation, and inducing plant defenses with a focus on technologies that have been practically used for disease and pest control in crop plants.

1.2 Stem Rust of Wheat

Stem rust (*Puccinia graminis* subsp. *graminis* Pers.: Pers.) has been devastating to wheat through many decades of production, especially during the 1950s in the United States. There are a number of reasons why it is less damaging now than it was previously, and this includes finding sources of stable resistance, reducing potential races in the population through the barberry (*Berberis vulgaris* L.) eradication program, and monitoring of potential races of the pathogen through a global network (Kolmer, 1996).

There are over 50 distinct genes for race-specific complete resistance to stem rust identified in wheat or in wild relatives of wheat, although not all of these resistance genes are broadly effective over space and time (Dubin and Rajaram, 1996). However, a few genes like *Sr31*, a gene originating from rye (*Secale cereale* L.) and transferred into wheat, was found to be very effective and was widely used, causing stem rust to decline to insignificant levels nearly everywhere in the world by the mid-1990s. In 1999, the resistance of *Sr31* was not effective to a new race, now known as UG99, found in Uganda (Stone, 2010). This new race rapidly spread and caused major epidemics in Kenya and Ethiopia, and spread south to South Africa and north and east to Iran. Because of the potential threat of this race to the world production of wheat, a number of organizations became involved to establish the Borlaug Global Rust Initiative in 2008 (globalrust.org). This initiative developed to monitor wheat rusts throughout the world and to reduce the impact of wheat rusts on yield by developing and deploying new effective resistance to wheat stem rust races, primarily UG99. To effectively combat wheat stem rust, there are both short- and long-term breeding strategies in place. Short-term strategies include pyramiding single-gene resistance and combining that with more long-term efforts to produce lines with effective partial resistance, based on genes for quantitative resistance (Bernardo et al., 2013; Pietrusińska et al., 2013; Tyagi et al., 2014; Mallick et al., 2015).

1.3 Rice Blast of Paddy Rice

Improvement of rice through conventional breeding techniques has made substantial progress and has been reviewed elsewhere (Guimaraes, 2009). One disease that has been a main focus of traditional breeding programs is the development of rice blast (*Magnaporthe oryzae* Cavara)-resistant cultivars. The fungus has a history of developing races, often resulting in rendering the resistance genes ineffective (Miah et al., 2013; Scheuermann et al., 2012). This is very similar to the wheat stem-rust situation and many other diseases of crops. The function of many of the rice resistance genes (over 100 loci) to control *M. oryzae* was described, and over 350 quantitative trait loci (QTL) for resistance to blast have been mapped in the rice genome, and blast resistant genes have been introgressed into lines through marker-assisted selection (Miah et al., 2013; Tanweer et al., 2015). Even though there have been concerted efforts and some success in breeding for resistance, the fungus has been resilient in causing blast and economic hardships (Scheuermann et al., 2012). In addition, it has been shown that pyramiding QTL alleles, each controlling a different response to *M. oryzae*, provides a non–race-specific, stable resistance to blast disease (Fukuoka et al., 2015), and thus, if it stands the test of time, would avoid the evolutionary "arms race" between rice breeders and *M. oryzae*. Rice, like other crops, has benefitted from transformation technology to engineer resistance against *M. oryzae* and to improve nutritional quality (accumulation of provitamin A and essential amino acids in endosperm) in basmati rice (Tanweer et al., 2015; Miah et al., 2013).

1.4 Corn Rootworm

There are a number of insects that cause considerable damage to maize. Often these insects vary by regional and seasonal differences. Corn rootworms (*Diabrotica* spp.) are considered a major pest throughout much of the corn-growing region in the United States and especially in the North Central region. The northern corn rootworm (*D. barberi* Smith and Lawrence) and the western corn rootworm (*D. virgifera virgifera* Le Conte) were estimated to result in annual yield losses and control costs exceeding $1 billion per year (Metcalf, 1986).

Traditionally, rootworms were managed through rotations and insecticidal seed or furrow treatments, and the rootworms complete their life cycle almost exclusively in fields where maize was consecutively grown for two years (Gassmann et al., 2012). In 1984 and in 1996, respectively, an extended diapause variant of the northern corn rootworm adapted to 2-year corn rotations, and a soybean variant of the western corn rootworm adapted its egg-laying behavior to lay eggs in crops other than corn (Levine and Oloumi-Sadeghi, 1996; Krysan et al., 1984). In 2000, the first new corn rootworm-resistant transgenic corn technology was introduced (Monsanto Yieldgard® Rootworm), which provided first-year benefits of about $138 million with only a 30% adoption rate (Alston et al., 2002). The production of maize hybrids able to produce different forms of Bt toxins was not long lived, as Bt-resistant corn rootworm populations became established in 2009 (Gassmann et al., 2012).

In 2015, the U.S. Environmental Protection Agency proposed a framework to prevent corn rootworm resistance from spreading by asking for ideas to delay Bt resistant corn rootworm pest populations (www.epa.gov/oppfead1/cb/csb_page/updates/2015/corn-rootworm-news.html). The proposed framework would change the way farmers use Bt corn, in order to slow the development of Bt-resistance. The goal is to prolong the durability and effectiveness of Bt corn to control the corn rootworm. In those areas deemed to be at risk of corn rootworm Bt-resistance, specific requirements of the manufacturers of Bt corn include use of crop rotation and hybrids with more than one Bt toxin, and that improvements be made in detection, sampling, and reliability of reporting Bt corn toxin resistance.

1.5 Leaf Rust of Soybean

The pathogen causing soybean leaf rust, *Phakopsora pachyrhizi* Syd., was first described in Japan in 1902 (Hennings, 1903), although a report from Australia predates this first description of the fungus when soybean rust was reported on soybeans in 1893 (Valder, 1893). Over the next century, the fungus was reported on soybeans and their allies throughout the Eastern Hemisphere, including Australia, China, India, Indonesia, Japan, Taiwan, and other countries in tropical and subtropical regions (Hartman et al., 2015). By 1994, *P. pachyrhizi* had spread beyond the Eastern Hemisphere and into Hawaii, and just a few years later, it was reported in many countries in Africa and South America (Hartman et al., 2015). In 2004, the disease was discovered on the North American continent in Louisiana (Schneider et al., 2005).

Somewhat similar to the wake-up call of stem rust race Ug99 on wheat, the concern of finding rust in the continental United States prompted a United States Department of Agriculture (USDA) document entitled Strategic Plan to Minimize the Impact of the Introduction and Establishment of Soybean Rust in Soybean Production in the United States (Anonymous, 2004). Preceding this document, there were several research projects headed by scientists based in the United States but working outside the country that were assessing fungicide efficacy, soybean rust resistance, aspects of fungal biology, and disease epidemiology (Isard et al., 2006; Miles et al., 2006, 2007).

Like the other pathogens on hosts previously mentioned, *P. pachyrhizi* has host-specific strains that produce different reaction types on soybean genotypes with differential resistance genes. This has been well documented in a number of studies and summarized in recent reviews (Hartman et al., 2005, 2011; Kelly et al., 2015). Although no formal race designations have been used to classify the isolates, a virulence formula designating which soybean genes are ineffective to a particular isolate would indicate the virulence of a purified isolate.

There are voluminous publications on soybean resistance to *P. pachyrhizi*. Since the introduction of the pathogen into the United States in 2004, concerted efforts have been made to find additional sources of resistance. The largest effort was through the USDA, when over 17,000 accessions, mostly PIs from the USDA Soybean Germplasm Collection, were evaluated for resistance (Miles et al., 2006). This and other work lead to the development and release of newer soybean rust-resistant germplasm (Boerma et al., 2011; Diers et al., 2013; Paul et al., 2010). Commercially available soybean cultivars with soybean rust resistance are not currently available in the United States, although these early releases and more newly developed resistance breeding lines will lead to cultivar releases in the near future with the possible deployment of soybean cultivars with stacked *Rpp* genes. A recent publication has summarized the known soybean rust resistance genes, which include loci *Rpp1* through *Rpp6* (Kelly et al., 2015). Soybean cultivars with resistance to *P. pachyrhizi* would add to a sustainable management program that now primarily includes the use of fungicide applications. Non-major gene resistance breeding for soybean rust management was reviewed in several other papers (Hartman et al., 2005; Miles et al., 2011), but their deployment in commercial soybean cultivars may be limited, as there has been little advancement made in breeding for nonspecific resistance, unlike wheat stem rust and rice blast.

2. Current Approaches to Increase Resistance Durability

2.1 Resistance Gene Pyramiding

Several decades ago, Johnson (1981) criticized the widely-held concept of durable resistance, which was resistance that remained effective in environments favorable for disease development, because of the lack of information on its mode of action, especially regarding host–pathogen interactions. He recognized that resistance durability could be improved by combining appropriate resistance genes into a crop host genotype that together control broad-spectrum resistance against pathogen populations with wide virulence potential. He also realized it was difficult to select for host genotypes with combinations of resistance genes using the technology available to breeders during that time without complex, costly resistance phenotyping and progeny testing using pest or pathogen populations with diverse virulence each generation during cultivar development.

Over the last couple of decades, with the development of low-cost, high-throughput molecular marker platforms, enabling marker-assisted selection (MAS) breeding procedures (Collard and Mackill, 2008) (see Chapter 4), advances in improvement in durability of genetic resistance in crops against pathogens and pests has been significantly facilitated (Joshi and Nayak, 2010). Using markers that closely flank regions that have high linkage disequilibrium (LD) with resistance genes in a MAS program, plants with the target marker alleles inherited from a resistant parent can now be readily selected during early (F_2) and subsequent generations of cultivar development, without the need for resistance phenotyping. The basic workflow involved in gene pyramiding using MAS is similar in all crops (Figure 2); however, many variations of this process have been practiced. The higher the LD between the marker and resistance gene, the lower the probability of recombination events separating the resistant parent marker alleles from the resistance gene, and the higher the confidence in MAS.

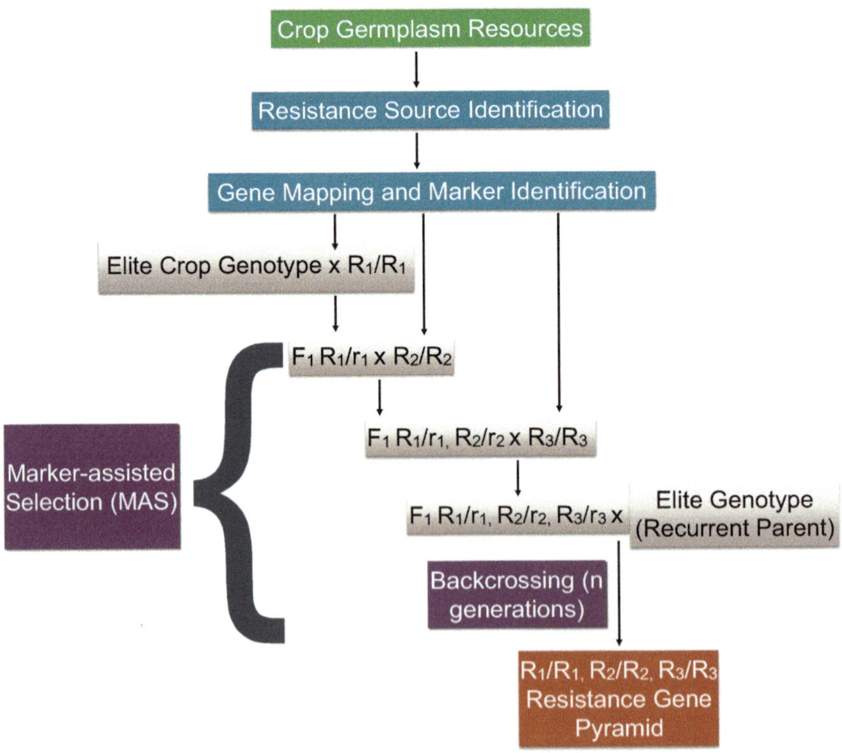

Figure 2 Illustration of the basic workflow used in producing resistance gene pyramids using marker-assisted selection (MAS).

This enables multiple independently inherited, non-allelic resistance genes to be combined together into a single cultivar using backcrossing (Collard and Mackill, 2008) or pedigree (Servin et al., 2004) breeding procedures to transfer genes from potentially several different resistant source parental genotypes (Joshi and Nayak, 2010). Therefore, the time to develop durable resistant cultivars with multiresistance genes was shortened considerably.

The basis for resistance durability is still not well understood (Mundt, 2014; Palloix et al., 2009). It is widely assumed that pathogen or pest population adaptation is slower on plants with multiple resistance genes compared with genotypes with single genes because of the time needed for virulence mutations and genetic recombination to occur (Collard and Mackill, 2008; Adugna, 2004), especially for pathogens without a sexual cycle (McDonald and Linde, 2002; Burdon, 1993). Based on this assumption, several resistance gene pyramids have been successfully developed using MAS and deployed in crop production.

Gene pyramids are multiple genes controlling a trait, such as pathogen or pest resistance, that are accumulated, combined, or stacked into a single genotype (Pedersen and Leath, 1988). Resistance gene pyramids usually consist of major, virulence-specific genes, but can include minor genes, defeated genes, effective genes, ineffective genes, race-specific genes, non–virulence-specific genes, or any other host genes involved in a resistance response to attack. This approach to increase resistance durability has been accomplished using traditional breeding and selection techniques. For instance, the best-known durable resistance using resistance gene pyramids may be the control of wheat leaf rust (*Puccinia recondita* Roberge ex Desm.), which was limited for decades in the United States with resistance gene pyramids (Schafer and Roelfs, 1985; Samborski, 1985), particularly with the two adult-plant resistance genes *Lr13* and *Lr34* in combination (Kolmer, 1996). Additional examples of the use of the resistance gene pyramiding approach using MAS to improve durability against diseases include *Soybean mosaic virus* (SMV) (Shi et al., 2009; Maroof et al., 2008) in soybeans, bacterial blight (*Xanthomonas oryzae* pv. *oryzae*) in rice (Sundaram et al., 2008), and root-knot nematode (*Meloidogyne arenaria* (Neal) Chitwood race 1) in peanuts (*Arachis hypogaea* L.) (Chu et al., 2011); and pests, including soybean aphid (*Aphis glycines* Matsumura) (McCarville and O'Neal, 2012; McCarville et al., 2014), wheat Hessian fly (*Mayetiola destructor* (Say)) (Sardesai et al., 2005), and wheat greenbug (*Schizaphis graminum* (Rondani)) (Porter et al., 2000). These examples are non-exhaustive, as there are numerous other examples of resistance gene pyramids that have been developed (Ramamoorthy et al., 2001) to broaden the spectrum to extend the effective life of host resistance against herbivores and plant pathogens of all kinds (Burdon, 1993; Castro et al., 2003; Joshi and Nayak, 2010; McDonald and Linde, 2002; Mundt, 2014; Suh et al., 2013; Tar'an et al., 2003, Tyagi et al., 2014). Pyramids of quantitative trait loci (QTLs), controlling quantitative resistance, have also been developed using MAS to improve durability (Castro et al., 2003; Tyagi et al., 2014).

Since the mechanism governing the durability of resistance gene pyramids is not understood and probably not universal (Collard and Mackill, 2008), it is still not known if the best resistance gene pyramiding strategy is to increase the numbers of qualitative or quantitative resistance genes in a pyramid, or select the best combinations of different resistance genes (Mundt, 2014). A recent study found diminishing efficacy with increasing numbers of quantitative resistance genes controlling partial resistance to foliar diseases in a wheat pyramid (Grimmer et al., 2014). Combining both strategies might be most effective, depending on the host–pathogen/pest system.

Advancements in molecular genetics and transformation techniques have enabled the pyramiding of transgenes to improve the durability of transgenic resistance (Joshi and Nayak, 2010; Mundt, 2014). The emergence of insect resistance to *Bacillus thuringiensis* (Bt) delta endotoxins has prompted the development of cotton and other crops with multiple Bt *Cry* endotoxin gene pyramids, with the hope that the use of pyramids will delay the selection of Bt endotoxin-resistant insect populations (Joshi and Nayak, 2010). However, theoretical and empirical evidence suggests that pyramiding Bt endotoxin genes will not provide sustainable control of target insects without integrated pest management (IPM) with other management practices that include cultural, chemical, and biological controls (Manyangarirwa et al., 2006). Development of synthetic meiotic-transferable mini-chromosomes to deliver pyramided transgenes or cloned natural host resistance genes into elite crop genotypes may offer a more rapid alternative to MAS to pyramid resistance genes in the future (Yu et al., 2007; Moose and Mumm, 2008).

2.2 Intelligent Deployment of Resistance Genes

Crop scientists have recognized the need to incorporate host resistance with other pathogen and pest management methods using IPM to increase resistance durability (Pink, 2002; Mundt, 2014). Cultural IPM methods that could increase durability include resistance gene rotation, crop seed mixtures of different resistant genotypes, increasing cropping diversity within the local and regional landscapes, and refugia using susceptible crop genotypes to reduce selection pressure toward resistance gene adaptation. IPM practices can be combined with intelligent deployment of resistance genes, based on knowledge of virulence potential, to further minimize pathogen/pest adaptation to resistance genes. Australian breeders have used knowledge gained from routine surveillance of wheat rust populations to practice "anticipatory resistance breeding" using a coordinated national effort, which has successfully reduced rust populations and virulence variability (McIntosh and Brown, 1997). With accelerating advancements in gene sequencing technologies enabling rapid identification and isolation of virulence genes in pathogen and pest populations, local and regional virulence potential in a pest or pathogen population can be more effectively surveyed and quickly assessed using PCR-based screening of virulence factors in samples of pathogens or pests. Information on virulence potential can be used to deploy host resistance genes, based on virulence intelligence to maximize their effectiveness and extend their usefulness (Hill et al., 2012). For this to work, a coordinated effort between seed companies, university and extension research scientists, crop consultants, and crop producers (Figure 3), similar to the national effort in Australia, is required to determine virulence potential and apply that knowledge to

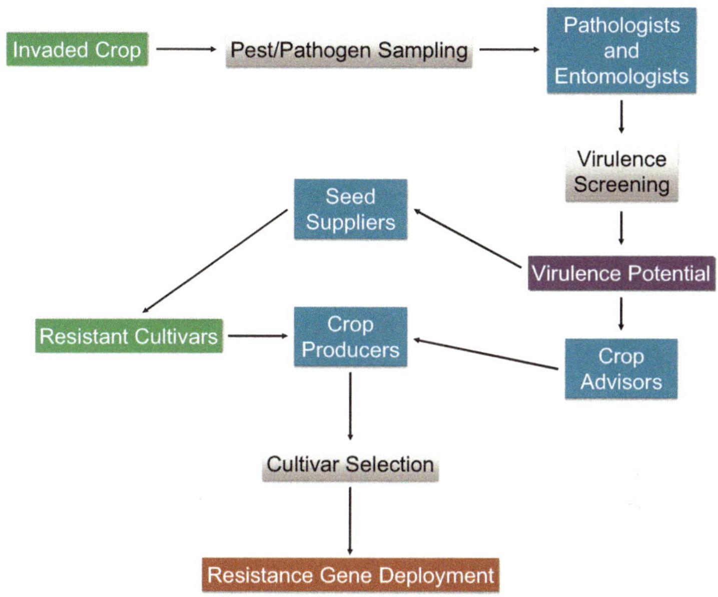

Figure 3 Diagram illustrating a coordinated effort to apply intelligent deployment of resistance genes.

guide the deployment of genes in space, across farms in a region, and time, to compensate for the dynamics of virulence across years, primarily caused by the selection imposed on pest or pathogen populations by host resistance (Adugna, 2004).

3. Improvement of Crop Resistance with Genetic Transformation

Transformation is the genetic gain of exogenous DNA (deoxyribonucleic acid) from a naturally occurring source or an engineered source into cells of bacteria, yeasts, plants, and animals. The discovery of plant transformation was tightly associated with the understanding of *Agrobacteria tumefaciens*, a bacterial pathogen that induces tumor-like overgrowth on susceptible plants (Kado, 2014). During the 1940s, observation of bacteria-free tumors promoted the idea that unknown substances may be transferred from pathogen into plants to induce tumors, and intensive studies eventually identified the unknown substance as a fragment of DNA (T-DNA), which carries oncogenic genes that move from a bacterial plasmid DNA, the tumor-inducing plasmid (Ti Plasmid), into plant cells during infection (Păcurar et al., 2011). The discovery stimulated the idea of transgenic plant cells with an exogenous gene to acquire a desired phenotype. After the first success in transgenic tobacco at 1983, numerous transgenic crops have been designed and developed to enhance resistance to pathogens and pests, some having a major impact in commercial production for the management of pathogens and pests (Dunwell, 2014).

3.1 Genetic Transformation Against Pests: Bt Cotton and Bt Corn

One of the most successful uses of transgenic crops to manage pests was the development of Bt cotton and Bt corn. These transgenic crops constitutively express exogenous endotoxin genes derived from Bt. Endotoxins are insecticidal crystal inclusions classified into several types; one of which is the largest 3d-Cry protein that is used in commercial products. Although the evolution of various 3d-Cry proteins from different Bt strains may be diverse, the toxicity mechanism is similar. In general, 3d-Cry proteins are activated after insect feeding, and after a complicated binding interaction to insect receptors on the gut membrane, 3d-Cry protein develops into an oligomeric complex that inserts into the insect membrane and creates pore formation in susceptible insects (Bravo et al., 2011). In addition to Bt, insecticidal toxins have been found in other organisms such as *Serratia entomophila* and *B. sphaericus*. These insecticidal toxins account for about 2% of the insecticide market as spray products, but are more commonly used by expressing endotoxins in crops through transformation (Pardo-López et al., 2013). After the first generation of Bt crops released into the market in 1996, which contained a single endotoxin against insects (Head and Greenplate, 2012), Bt crops have been shown to stabilize yield (Edgerton et al., 2012) and reduce the application of chemical insecticides (Lu et al., 2012). About 67% and 30% of the maize production is transgenic in the United States and globally, respectively, and more than 70% of cotton production is transgenic (James, 2014; Tabashnik et al., 2013). Yield benefits of growing Bt corn range from 11% to 66% yield increase under the most destructive rootworm attack (Haegele and Below, 2013; Ma et al., 2009), and in cotton, a yield increase of 24% was reported for growing Bt cotton in India (Kathage and Qaim, 2012). Moreover, reduction of insect damage on Bt corn has been shown to reduce mycotoxin-producing fungal pathogen infection. Up to 75% of mycotoxin reduction has been observed, which provides an additional advantage of using Bt corn to control insect damage (Munkvold and Hellmich, 1999). Although the promise of using Bt crops has been well documented, concerns about insect adaptation to the Bt endotoxins continue, and indeed, several cases report on field-evolved Bt resistance in pests worldwide. It has been suggested that field-selected resistance may occur within 2 years, but it is possible to delay Bt resistance by 15 years by ensuring that the high-dose Bt endotoxin is effective on the target pest and by evaluating the frequency of Bt-resistant alleles in a pest population in advance of commercializing a new Bt crop (Tabashnik et al., 2013). In addition, refuge strategy has been shown to be effective in delaying the outbreak of a Bt-resistant population of pests. Refugia maintain non-Bt crops nearby Bt crops to reduce selection pressure on Bt-resistant mutations, which are generally recessive. Nevertheless, refuge strategy depends on both the high-dose assumption and the frequency of Bt-resistant alleles in a pest population. Without completely understanding these assumptions, it may be inferred that Bt-resistance preexisted in a population before applying Bt, and then a Bt-resistant population developed due to selection pressure imposed by the Bt-transformed crop. For example, the resistance of *Helicoverpa zea* to Bt cotton was once considered a field-evolved resistance, but other evidence suggested the possibility of preexisting resistance in the *H. zea* population (Luttrell and Jackson, 2012). Other cases of Bt resistance that evolved in the field include resistant *Busseola fusca* to Bt corn that had Cry1Ab reported in South Africa, resistant *Pectinophora gossypiella* to Bt cotton that had Cry1Ac reported in India, and resistant *Spodoptera frugiperda* to Bt corn that had Cry1F reported in the United States. All of those cases emerged on first generation Bt crops expressing a single Bt endotoxin. Recently, commercial second generation Bt-resistant crops were developed by pyramiding two or more endotoxin genes into a cultivar. With appropriate prescreening for Bt-resistant frequency and their tolerance to BT endotoxins in a pest population, accurate refuge strategy, and using pyramids with more than one Bt toxin, Bt-susceptible crop pests may be sustainably managed into the future.

3.2 Genetic Transformation against Pathogens: Papaya Ringspot Virus

A successful case using genetic transformation to improve resistance against a disease is the case of transgenic papaya (*Carica papaya*) resistant to *Papaya ringspot virus* (PRSV). PRSV is a potyvirus that is primarily transmitted by aphids. The disease has been reported worldwide, including in Hawaii, where PRSV is the major disease of papayas. Among several viral symptoms that PRSV induces on infected plants, the reduction of sugar content resulting in bitter taste makes the fruit unmarketable (Tripathi et al., 2008). It has been shown that controlling the aphid vector was not effective alone in controlling PRSV (Pernezny and Litz, 2003). Natural resistance in *C. papaya* has been reported to be ineffective in cultivated papaya (Ziebell and Carr, 2010). Although resistance was found in some closely related *Carica* species, sexual incompatibility increased the challenges for conventional breeding for PRSV resistance (Gonsalves et al., 2006). A management approach utilizing the phenomenon of cross-protection, which involves infecting plants with a mild strain of virus that interferes with subsequent infection of a more severe virus strain, has been applied to control virus diseases. This approach was first investigated from the 1950s to 1980s on cocoa to fight against cocoa swollen shoot virus. Commercial application of cross-protection was applied to *Citrus tristeza virus, Zucchini yellow mosaic virus*, and PRSV (Ziebell and Carr, 2010). In the absence of natural mild strains of PRSV, some mild strains of PRSV, such as PRSV-HA 5-1, was generated from a severe strain of PRSV-HA by chemical mutagenesis. PRSV-HA 5-1 has been applied to control PRSV in field trials by massive inoculation of mild strains. However, these early attempts to use cross-protection were not very successful. One of the limitations was that mild strains were only effective against closely related strains or parental strains (Yeh and Gonsalves, 1984). It was suggested the protectant mild strains should be generated based on endemic virulent strains. Environmental variability in the expression of cross protection remains a challenge as high disease pressure and cold temperatures speed up the breakdown of cross-protection (Gonsalves, 1998). Other problems encountered during early use of cross-protection included incomplete inoculation of mild strains, differential responses of mild strains on different papaya cultivars, as well as the possibility of spread of mild strains to other plant hosts (Ziebell and Carr, 2010).

The mechanism of cross-protection is not understood, and more than one mechanism may control the effect of cross-protection (Kung et al., 2014). One of the most accepted theories on the mechanism was based on RNA (ribonucleic acid) silencing, which led to the development of transgenic papayas using RNA silencing to control the disease. The effectiveness of transgenic RNA silencing depends on the specificity of transgene sequences and the geographic diversity of PRSV (Tennant et al., 1994). Broad-spectrum resistant transgenic papayas, based on homologs of viral targets, have been developed (Bau et al., 2003) and applied in the field (Azad et al., 2014). Transgenic papayas have been used in more than 70% of the papaya production in Hawaii (Gonsalves et al., 2006), and successfully rescued the industry from a PRSV-caused demise. The approach has been considered as a long-term protection for managing PRSV (Azad et al., 2014; Fermin et al., 2010). Moreover, in addition to managing PRSV, an RNA silencing-based transgenic approach has been proposed to develop a brand new era of pest-resistant crops (Baum et al., 2007; Gordon and Waterhouse, 2007). Intense studies have also been focused on evaluating the potential of this strategy on pest management (Terenius et al., 2011). Looking back from the 1940s to the 2000s, the advance of knowledge of plant biology and technologies on plant transformation gradually made novel ideas practical to sustain yield production by managing pests and diseases. Although there is great promise for the use of advanced plant biotechnology to improve crop resistance, deployment of transgenic resistant crops is currently being constrained by government regulation and societal nonacceptance.

4. Inducing Plant Defenses

For over 100 years, it has been well known that plants can be preconditioned to increase resistance against a broad range of diseases, a phenomenon initially known as physiological acquired immunity (Chester, 1933). Since then, scientists have been investigating the mechanisms behind this preconditioning. In 1961, the term "systemic acquired resistance" was coined when it was demonstrated that infecting tobacco leaves with *Tobacco mosaic virus* (TMV) increased resistance in the plant to later TMV infections, as well as *Tobacco necrosis virus* and other bacterial pathogens (Ross, 1961a,b). Since then, a substantial amount of research has been done to further understand different forms of induced resistance. Inducing resistance is still not fully understood, but has been shown to be effective in a broad range of crops against bacterial, fungal, and viral pathogens, as well as pests.

4.1 *Systemic Acquired Resistance*

Systemic acquired resistance (SAR) is a form of induced resistance that is activated throughout a plant after being exposed to elicitors from virulent, avirulent, or nonpathogenic microbes, or artificial chemical stimuli such as chitosan or salicylic acid (SA) (Figure 4; Vallad and Goodman, 2004; Gozzo and Faoro, 2013). Induction of SAR is characterized by the accumulation of salicylic acid to stimulate defense

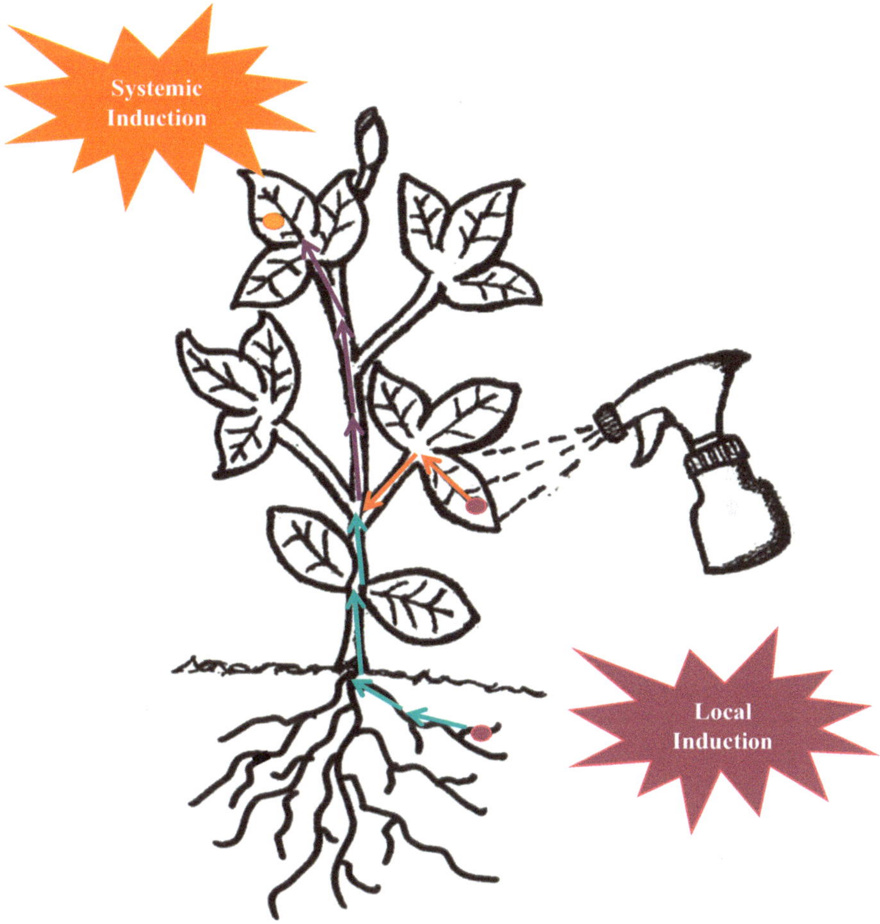

Figure 4 Simple schematic representation of how local induced resistance triggered by exogenous application of chemical elicitors (red (dark gray in print versions) arrows) or microbial communities in the soil (blue (light gray in print versions) arrows) induce systemic resistance (purple (gray in print versions) arrows).

mechanisms, often resulting in a localized hypersensitive response. SA is not a mobile signal agent, but is proposed to be the first chemical in the induction of pathogenesis-related (*PR*) genes, especially *PR-1* (Sticher et al., 1997; Fu and Dong, 2013). *PR* genes code for proteins such as chitinases and β-1,3-glucanases among others, and play a role in either preventing or slowing colonization of pathogens in the host (Sudisha et al., 2012).

Induction of SAR has increased disease resistance to a number of pathogens and pests, including *Ascochyta fabae* and *Uromyces viciae-fabae* on faba beans and *Meloidogyne incognita* and *Ralstonia solanacearum* on tomatoes (Molinari and Baser, 2010; Pradhanang et al., 2005; Sillero et al., 2012). It has been suggested that SAR is most effective against biotrophic and hemibiotrophic pathogens and not against necrotrophic pathogens since it stimulates the SA pathway (Glazebrook, 2005; Oliver and Ipcho, 2004). Repeated inoculations can increase the level of resistance, but it can take several days for SAR to develop throughout the host plant (Kuc, 1982).

4.2 Induced Systemic Resistance

Induced systemic resistance (ISR) is activated similarly to SAR by chemical elicitors, but is usually linked to stimulation by plant growth-promoting rhizobacteria (PGPR) and beneficial fungi (Figure 5; Pieterse and Van Wees, 2015). ISR also differs from SAR because it is independent of the SA pathway and does not activate *PR* genes or exhibit an HR response, but in some cases, it may stimulate production of siderophores or SA, similar to SAR (De Vleesschauwer and Hofte, 2009; Pieterse and Van Wees, 2015). Phytohormones jasmonic acid (JA) and ethylene (ET) are necessary for activation of ISR, which was affirmed when *Arabidopsis thaliana* JA and ET response mutants did not show increased resistance against *Pseudomonas syringae* pv. *tomato* after colonization of PGPR bacteria (De Vleesschauwer and Hofte, 2009). Activation does not upregulate JA- and ET-regulated genes, but increases responsiveness of

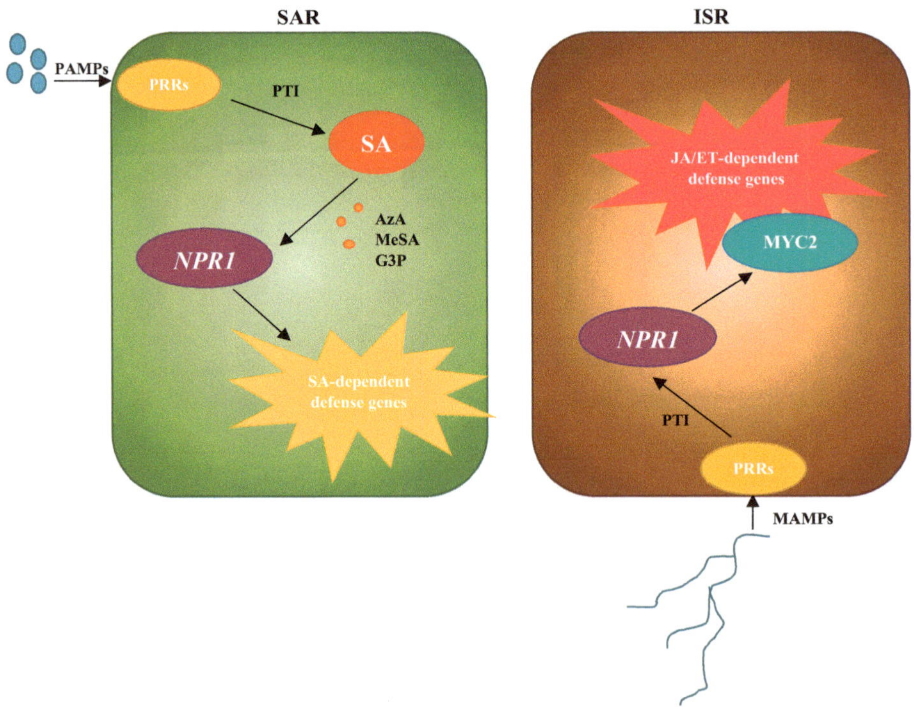

Figure 5 Schematic representation of the molecular mechanisms involved in activation of SAR and ISR. PAMP, pathogen-associated molecular pattern; PRR, pattern-recognition receptor; PTI, PAMP-triggered immunity; SA, salicylic acid; AzA, azelaic acid; MeSA, methyl ester of SA; G3P, glycerol-3-phosphate; *NPR1*, natriuretic peptide receptor 1; MAMP, microbe-associated molecular pattern; JA, jasmonic acid; ET, ethylene.

these pathways to elicitation, which is proposed to be mediated by the transcription factor MYC2 (Kazan and Manners, 2013). This phenomenon is referred to as priming (Conrath et al., 2002). The ability to prime plants allows inducing ISR to be more effective against necrotrophic pathogens than biotrophic or hemibiotrophic pathogens since it does not induce a hypersensitive response (Glazebrook, 2005; van Loon et al., 1998). The application of *Bacillus* spp. on rice protected the plant from bacterial leaf blight (Udayashankar et al., 2011). *Trichoderma* spp. have also been shown to induce ISR, increasing resistance to bacterial, fungal, and viral pathogens on a number of hosts, including *Botrytis cinerea* on beans, lettuce, peppers, and tomatoes, also *Colletotrichum orbiculare,* and *P. syringae* pv. *lachrymans* on cucumber, *Phytophthora capsici* on peppers, and *Cucumber green mosaic virus* on tomatoes (Ahmed et al., 2000; Koike et al., 2001; De Meyer et al., 1998; Yedidia et al., 2003; Vitti et al., 2014).

4.3 Eliciting Compounds

SA was the first chemical shown to elicit a defense response when exogenous SA was used to induce resistance in tobacco to TMV (White, 1979). Although there are a good number of studies showing the efficacy of SA, its analogs, such as benzo(1,2,3) thiadiazole-7-carbothioic acid *S*-methyl ester (BTH), also known as acibenzolar-*S*-methyl, have been shown to be substantially more effective. Application of BTH stimulates accumulation of mitogen-activated protein kinase 3, which stimulates defense pathways (Beckers and Conrath, 2007). In *A. thaliana*, BTH induced *PR-1* messenger RNA (mRNA) accumulation in ethylene response and jasmonate-insensitive mutants, suggesting that BTH acted independently of those pathways. *PR-1* was also expressed in transgenic plants expressing the salicylate hydroxylase gene, *NahG*, suggesting BTH activates resistance downstream of SA accumulation (Gozzo and Faoro, 2013). BTH is the most well-studied elicitor, showing effectiveness against a wide range of pathogens and pests, but the commercialized product was short-lived because of variable effectiveness (Gozzo and Faoro, 2013).

Chitin is the main component of fungal cell walls and is found in the exoskeleton of crustaceans (Iriti and Varoni, 2015). Chitin and its derivative chitosan have been well-studied for their antimicrobial activity against bacteria and fungi, but have also been shown to induce SAR in plants. Exogenous application of chitosan generates an accumulation of phenolics and callose, and has been shown to reduce diseases caused by bacterial, fungal, and viral pathogens. As with BTH, the efficacy of chitosan is dependent on the plant and pathogen, and as an example of its effectiveness, it has been shown to reduce the severity of powdery mildew of grape under high disease pressure (Gozzo and Faoro, 2013; Iriti and Varoni, 2015). In 2000, a commercialized product that contained 4% chitosan was advertised as a disease resistance activator, but the product has not received widespread popularity because the results are variable.

Multiple comprehensive reviews of PGPR-mediated ISR have shown many strains of *Bacillus, Pseudomonas*, and *Serratia* were able to induce resistance on a wide range of hosts (De Vleesschauwer and Hofte, 2009). Of the effective bacterial strains found to reduce disease, *Bacillus pumilus* INR-7 has shown remarkable results in field trials against angular leaf spot, anthracnose, bacterial wilt, and cucumber beetles when applied to cucumbers; however, limited success against viral pathogens has been shown (Raupach and Kloepper, 1998, 2000; Wei et al., 1996; Zehnder et al., 2001).

4.4 Impact of Inducers

The concept of induced resistance has been widely accepted, but its role as an effective disease-management strategy is still debatable. Supporters of the use of resistance elicitors focus on their ability to reduce the need for chemical pesticides; however, this is a complex issue. Along with efficacy in disease management, chemical plant defense inducers have been evaluated for their impact on other aspects of production, including yield. A review compiled previous studies on the costs and benefits of inducing resistance in plants (Cipollini and Heil, 2010). There have been numerous studies showing the benefits of inducing resistance with chemical elicitors to reduce disease; however, this does not always result in an increase in yield and could potentially have adverse effects (Cole, 1999; Louws et al., 2001; Romero et al., 2001). The lack of knowledge about induced resistance and the reluctance to incorporate the concept into existing models of disease management theory has delayed the use of this management strategy. Even with these costs, numerous studies have shown that the benefits of using these elicitors highly outweigh the costs, though this is dependent on a number of biotic and abiotic variables, such as environmental conditions and the level of disease pressure. Although use of elicitors alone has not gained much popularity, these disease-resistance agents are now frequently being added to pesticides to gain extra protection.

5. Future Considerations

The focus of this chapter has been upon how traditional breeding, molecular breeding, genetic engineering, and application of treatments that boost plant immune responses have had significant economic success in managing pathogens and pests that limited crop production with host resistance. Also emphasized was the fact that host resistance must be integrated with other management techniques to extend the useful life of resistance factors. Several novel management technologies are under current development.

Methods to sanitize pest-infested plants—for instance, ultrasonic or pneumatic machines (Vincent et al., 2003)—have been used for control of the wheat weevil (*Sitophilus granarius*) (Pradzynska, 1982) and Colorado potato beetle (*Leptinotarsa decemlineata*); however, it has been pointed out that the efficiency of aspiration requires improvements (Lacasse et al., 2001; Misener and Boiteau, 1993; Vincent and Boiteau, 2001). One of the limitations was the energy input for generating field-wide ultrasonic or pneumatic airflow, but with improvement in solar power–driven robots, it may be that mechanical devices in the future will end up benefiting agriculture.

The use has been studied of nanoparticles that contain Ag, Cu, and Zn applied to plants to suppress pathogen germination, growth, and infection has been studied. For example, Ag on grapheme oxide reduced bacterial spot of tomatoes, and reduces the colony-forming units in culture plates of two pathogenic fungi, *Bipolaris sorokiniana* and *Magnaporthe oryzae* (Jo et al., 2009; Ocsoy et al., 2013). Nanoparticles with Cu were shown to reduce the growth of *P. infestans*, and foliar applications may control soil-borne vascular pathogens such as *Fusarium* and *Verticillium* species, which indicates the movement of nanoparticle Cu may be phloem-dependent (Giannousi et al., 2013). Nanoparticles with Zn were shown to have potential to inhibit bacterial species (*Pseudomonas*), as well as fungal species of *Aspergillus*, *Fusarium*, *Mucor*, and *Rhizopus* (Servin et al., 2015). In addition, micronutrients such as Fe, Mn, and Zn may be more effectively applied through nanoparticles, which may be easily absorbed through stomata (Eichert et al., 2008). Although the use of nanoparticles may improve plant health and enhance disease resistance, details about adverse affects of nanoparticle technology need further research before nanotechnology will be used in improving crop health and productivity (Servin et al., 2015).

In addition to the above examples, "precision, smart, intelligence-driven farming" is the vision for future farming. The vision includes the use of more robotics, on the ground and in the air, to alleviate the impact of crop pests. This computer-based precision will push agriculture technologies and engineering practices to become more effective with a faster turnaround time from research into practice.

Two major challenges for the future of host resistance development are climate change and government regulation of agricultural technologies. Regarding climate change, with increasing global temperatures, adaptation of pests and pathogens to host resistance may be accelerated by shorter pest and pathogen life cycles (Petzoldt and Seaman, 2006). Therefore, the importance of efforts to improve resistance durability will continue to increase. Future advances in agricultural technology to increase or enhance host resistance will likely involve molecular modification of host plants and development of novel biological products, including microorganisms and metabolites. For the benefits of new advances to be realized in crop production, a regulatory balance between pre-commercial testing and post-market review must be found between technology developers and government regulators (Barrows et al., 2014).

References

Adugna, A., 2004. Alternate approaches in deploying genes for disease resistance in crop plants. Asian Journal of Plant Sciences 3, 618–623.

Ahmed, A.S., Sanchez, C., Candela, M., 2000. Evaluation of induction of systemic resistance in pepper plants (*Capsicum annuum*) to *Phytophthora capsici* using *Trichoderma harzianum* and its relation with capsidiol accumulation. European Journal of Plant Pathology 106, 817–824.

Alexandratos, N., Bruinsma, J., 2012. World agriculture towards 2030/2050: the 2012 revision. In: ESA Working Paper No. 12-03. Rome, FAO.

Alston, J., Hyde, J., Marra, M., Mitchell, P., 2002. An ex ante analysis of the benefits from the adoption of corn rootworm resistant transgenic corn technology. AgBio Forum 5, 71–84.

Anonymous, 2004. Strategic Plan to Minimize the Impact of the Introduction and Establishment of Soybean Rust in Soybean Production in the United States. U.S. Dept. of Agriculture, 31 pp.

Azad, M.A.K., Amin, L., Sidik, N.M., 2014. Gene technology for papaya ringspot virus disease management. The Scientific World Journal 2014.

Barrows, G., Sexton, S., Zilberman, D., 2014. Agricultural biotechnology: the promise and prospects of genetically modified crops. The Journal of Economic Perspectives 28, 99–119.
Bau, H.-J., Cheng, Y.-H., Yu, T.-A., Yang, J.-S., Yeh, S.-D., 2003. Broad-spectrum resistance to different geographic strains of papaya ringspot virus in coat protein gene transgenic papaya. Phytopathology 93, 112–120.
Baum, J.A., Bogaert, T., Clinton, W., Heck, G.R., Feldmann, P., Ilagan, O., Johnson, S., Plaetinck, G., Munyikwa, T., Pleau, M., 2007. Control of coleopteran insect pests through RNA interference. Nature Biotechnology 25, 1322–1326.
Beckers, G.J., Conrath, U., 2007. Priming for stress resistance: from the lab to the field. Current Opinion in Plant Biology 10, 425–431.
Bernardo, A., Bowden, R., Rouse, M., Newcomb, M., Marshall, D., Bai, G., 2013. Validation of molecular markers for new stem rust resistance genes in US hard winter wheat. Crop Science 53, 755–764.
Boerma, H.R., Monteros, M.J., Ha, B.-K., Wood, E., Phillips, D.V., Walker, D.R., Missaoui, A.M., 2011. Registration of Asian soybean rust-resistant soybean germplasm G01-PR16. Journal of Plant Registrations 5, 118–122.
Bravo, A., Likitvivatanavong, S., Gill, S.S., Soberón, M., 2011. *Bacillus thuringiensis*: a story of a successful bioinsecticide. Insect Biochemistry and Molecular Biology 41, 423–431.
Burdon, J., 1993. Genetic variation in pathogen populations and its implications for adaptation to host resistance. In: Jacobs, T., Parlevliet, J.E. (Eds.), Durability of Disease Resistance. Kluwer Academic Publishers.
Castro, A.J., Capettini, F., Corey, A., Filichkina, T., Hayes, P.M., Kleinhofs, A., Kudrna, D., Richardson, K., Sandoval-Islas, S., Rossi, C., 2003. Mapping and pyramiding of qualitative and quantitative resistance to stripe rust in barley. Theoretical and Applied Genetics 107, 922–930.
Chester, K.S., 1933. The problem of acquired physiological immunity in plants. The Quarterly Review of Biology 8, 275–324.
Chu, Y., Wu, C., Holbrook, C., Tillman, B., Person, G., Ozias-Akins, P., 2011. Marker-assisted selection to pyramid nematode resistance and the high oleic trait in peanut. Plant Genome 4, 110–117.
Cipollini, D., Heil, M., 2010. Costs and benefits of induced resistance to herbivores and pathogens in plants. CAB Reviews: Perspectives in Agriculture, Veterinary Science, Nutrition and Natural Resources 5.
Cole, D.L., 1999. The efficacy of acibenzolar-S-methyl, an inducer of systemic acquired resistance, against bacterial and fungal diseases of tobacco. Crop Protection 18, 267–273.
Collard, B., Mackill, D., 2008. Marker-assisted selection: an approach for precision plant breeding in the twenty-first century. Physiological Transactions of the Royal Society B: Biological Sciences 363, 557–572.
Conrath, U., Pieterse, C.M.J., Mauch-Mani, B., 2002. Priming in plant–pathogen interactions. Trends in Plant Science 7, 210–216.
De Meyer, G., Bigirimana, J., Elad, Y., Höfte, M., 1998. Induced systemic resistance in *Trichoderma harzianum* T39 biocontrol of *Botrytis cinerea*. European Journal of Plant Pathology 104, 279–286.
De Vleesschauwer, D., Hofte, M., 2009. Rhizobacteria-induced systemic resistance. In: Vanloon, L.C. (Ed.), Plant Innate Immunity. Academic Press Ltd-Elsevier Science Ltd, London.
Diers, B.W., Kim, K.S., Frederick, R.D., Hartman, G.L., Unfried, J.R., Schultz, S.J., Cary, T.R., 2013. Registration of eight soybean germplasm lines resistant to soybean rust. Journal of Plant Registrations 8, 96–101.
Dubin, H., Rajaram, S., 1996. Breeding disease-resistant wheats for tropical highlands and lowlands. Annual Review of Phytopathology 34, 503–526.

Dunwell, J.M., 2014. Transgenic cereals: current status and future prospects. Journal of Cereal Science 59, 419–434.
Edgerton, M.D., Fridgen, J., Anderson Jr., J.R., Ahlgrim, J., Criswell, M., Dhungana, P., Gocken, T., Li, Z., Mariappan, S., Pilcher, C.D., 2012. Transgenic insect resistance traits increase corn yield and yield stability. Nature Biotechnology 30, 493–496.
Eichert, T., Kurtz, A., Steiner, U., Goldbach, H.E., 2008. Size exclusion limits and lateral heterogeneity of the stomatal foliar uptake pathway for aqueous solutes and water–suspended nanoparticles. Physiologia Plantarum 134, 151–160.
Fermin, G.A., Castro, L.T., Tennant, P.F., 2010. CP-transgenic and non-transgenic approaches for the control of papaya ringspot: current situation and challenges. Transgenic Plant Journal 4, 1–15.
Fu, Z., Dong, X., 2013. Systemic acquired resistance: turning local infection into global defense. Annual Review of Plant Biology 64, 839–863.
Fukuoka, S., Saka, N., Mizukami, Y., Koga, H., Yamanouchi, U., Yoshioka, Y., Hayashi, N., Ebana, K., Mizobuchi, R., Yano, M., 2015. Gene pyramiding enhances durable blast disease resistance in rice. Scientific Reports 5.
Gassmann, A., Petzold-Maxwell, J., Keweshan, R., Dunbar, M., 2012. Western corn rootworm and Bt maize: challenges of pest resistance in the field. GM Crops & Food: Biotechnology in Agriculture and the Food Chain 3, 235–244.
Giannousi, K., Avramidis, I., Dendrinou-Samara, C., 2013. Synthesis, characterization and evaluation of copper based nanoparticles as agrochemicals against *Phytophthora infestans*. RSC Advances 3, 21743–21752.
Glazebrook, J., 2005. Contrasting mechanisms of defense against biotrophic and necrotrophic pathogens. Annual Review of Phytopathology 43 Palo Alto: Annual Reviews.
Gonsalves, D., 1998. Control of papaya ringspot virus in papaya: a case study. Annual Review of Phytopathology 36, 415–437.
Gonsalves, D., Vegas, A., Prasartsee, V., Drew, R., Suzuki, J., Tripathi, S., 2006. Developing papaya to control papaya ringspot virus by transgenic resistance, intergeneric hybridization, and tolerance breeding. Plant Breeding Reviews 26, 35–78.
Gordon, K.H., Waterhouse, P.M., 2007. RNAi for insect-proof plants. Nature Biotechnology 25, 1231–1232.
Gozzo, F., Faoro, F., 2013. Systemic acquired resistance (50 years after discovery): moving from the lab to the field. Journal of Agricultural and Food Chemistry 61, 12473–12491.
Grimmer, M., Boyd, L., Clarke, S., Paveley, N., 2014. Pyramiding of partial disease resistance genes has a predictable, but diminishing, benefit to efficacy. Plant Pathology 64, 748–753.
Guimaraes, E.P., 2009. Rice breeding. In: Carena, M. (Ed.), Cereals, the Banks and the Italian Economy. Springer Science and Business Media, LLC, pp. 99–126.
Haegele, J.W., Below, F.E., 2013. Transgenic corn rootworm protection increases grain yield and nitrogen use of maize. Crop Science 53, 585–594.
Hartman, G.L., Hill, C.B., Twizeyimana, M., Miles, M.R., Bandyopadhyay, R., 2011. Interaction of soybean and *Phakopsora pachyrhizi*, the cause of soybean rust. CAB Reviews: Perspectives in Agriculture, Veterinary Science, Nutrition and Natural Resources 6, 1–13.
Hartman, G.L., Miles, M.R., Frederick, R.D., 2005. Breeding for resistance to soybean rust. Plant Disease 89, 664–666.
Hartman, G.L., Sikora, E.F., Rupe, J.C., 2015. Rust. In: Hartman, G.L., Sikora, E.F., Rupe, J.C. (Eds.), Compendium of Soybean Diseases and Pests. American Phytopathological Society, St. Paul.
Head, G.P., Greenplate, J., 2012. The design and implementation of insect resistance management programs for Bt crops. GM Crops & Food 3, 144–153.

Hennings, V.P., 1903. A few new Japanese Uredinaceae. Hedwigia 42, S107–S108.
Hill, C., Chirumamilla, A., Hartman, G., 2012. Resistance and virulence in the soybean-*Aphis glycines* interaction. Euphytica 186, 635–646.
Iriti, M., Varoni, E., 2015. Chitosan-induced antiviral activity and innate immunity in plants. Environmental Science and Pollution Research 22, 2935–2944.
Isard, S.A., Dufault, N.S., Miles, M.R., Hartman, G.L., Russo, J.M., De Wolf, E.D., Morel, W., 2006. The effect of solar irradiance on the mortality of *Phakopsora pachyrhizi* urediniospores. Plant Disease 90, 941–945.
James, C., 2014. Global Status of Commercialized Biotech/GM Crops in 2014. Pocket K No. 16.
Jo, Y.-K., Kim, B.H., Jung, G., 2009. Antifungal activity of silver ions and nanoparticles on phytopathogenic fungi. Plant Disease 93, 1037–1043.
Johnson, R., 1981. Durable resistance: definition of, genetic control, and attainment in plant breeding. Phytopathology 71, 587–588.
Joshi, R.K., Nayak, S., 2010. Gene pyramiding-A broad spectrum technique for developing durable stress resistance in crops. Biotechnology and Molecular Biology Review 5, 51–60.
Kado, C., 2014. Historical account on gaining insights on the mechanism of crown gall tumorigenesis by *Agrobacterium tumefaciens*. Frontiers in Microbiology. http://dx.doi.org/10.3389/fmicb.2014.00340.
Kathage, J., Qaim, M., 2012. Economic impacts and impact dynamics of Bt (*Bacillus thuringiensis*) cotton in India. Proceedings of the National Academy of Sciences United States of America 109, 11652–11656.
Kazan, K., Manners, J., 2013. MYC2: the master in action. Molecular Plant 6, 686–703.
Kelly, H., Dufault, N., Walker, D., Isard, S., Schneider, R., Giesler, L., Wright, D.L., Marios, J.J., Hartman, G., 2015. From select agent to an established pathogen, the response to *Phakopsora pachyrhizi* in North America. Phytopathology 105, 905–916.
Koike, N., Hyakumachi, M., Kageyama, K., Tsuyumu, S., Doke, N., 2001. Induction of systemic resistance in cucumber against several diseases by plant growth-promoting fungi: lignification and superoxide generation. European Journal of Plant Pathology 107, 523–533.
Kolmer, J., 1996. Genetics of resistance to wheat leaf rust. Annual Review of Phytopathology 34, 435–455.
Krysan, J., Jackson, J., Lew, A., 1984. Field termination of egg diapause in *Diabrotica* with new evidence of extended diapause in *D. barberi* (Coleoptera: Chrysomelidae). Environmental Entomology 13, 1237–1240.
Kuc, J., 1982. Induced immunity to plant disease. Bioscience 32, 854–860.
Kung, Y.-J., Lin, P.-C., Yeh, S.-D., Hong, S.-F., Chua, N.-H., Liu, L.-Y., Lin, C.-P., Huang, Y.-H., Wu, H.-W., Chen, C.-C., 2014. Genetic analyses of the FRNK motif function of *Turnip mosaic virus* uncover multiple and potentially interactive pathways of cross-protection. Molecular Plant-Microbe Interactions 27, 944–955.
Lacasse, B., Laguë, C., Roy, P.M., Khelifi, M., Bourassa, S., Cloutier, C., 2001. Pneumatic control of Colorado potato beetle. In: Vincent, C., Panneton, B., Fleurat-Lessard, F. (Eds.), Physical Control in Plant Protection, pp. 282–293.
Levine, E., Oloumi-Sadeghi, H., 1996. Western corn rootworm (Coleoptera: Chrysomelidae) larval injury to corn grown for seed production following soybeans grown for seed production. Journal of Economic Entomology 89, 1010–1016.
van Loon, L.C., Bakker, P., Pieterse, C.M.J., 1998. Systemic resistance induced by rhizosphere bacteria. Annual Review of Phytopathology 36, 453–483.
Louws, F.J., Wilson, M., Campbell, H.L., Cuppels, D.A., Jones, J.B., Shoemaker, P.B., Sahin, F., Miller, S.A., 2001. Field control of bacterial spot and bacterial speck of tomato using a plant activator. Plant Disease 85, 481–488.

Lu, Y., Wu, K., Jiang, Y., Guo, Y., Desneux, N., 2012. Widespread adoption of Bt cotton and insecticide decrease promotes biocontrol services. Nature 487.

Luttrell, R.G., Jackson, R.E., 2012. *Helicoverpa zea* and Bt cotton in the United States. GM Crops and Food: Biotechnology in Agriculture and the Food Chain 3, 213–227.

Ma, B., Meloche, F., Wei, L., 2009. Agronomic assessment of Bt trait and seed or soil-applied insecticides on the control of corn rootworm and yield. Field Crops Research 111, 189–196.

Mallick, N., Vinod, Sharma, J., Tomar, R., Sivasamy, M., Prabhu, K., 2015. Marker-assisted backcross breeding to combine multiple rust resistance in wheat. Plant Breeding 134, 172–177.

Manyangarirwa, W., Turnbull, M., Mccutcheon, G.S., Smith, J.P., 2006. Gene pyramiding as a Bt resistance management strategy: how sustainable is this strategy? African Journal of Biotechnology 5, 781–785.

Maroof, S., Jeong, S., Gunduz, I., Tucker, D., Buss, G., Tolin, S., 2008. Pyramiding of soybean mosaic virus resistance genes by marker-assisted selection. Crop Science 48, 517–526.

McCarville, M.T., O'neal, M., 2012. Measuring the benefit of biological control for single gene and pyramided host plant resistance for *Aphis glycines* (Hemiptera: Aphididae) management. Journal of Economic Entomology 105, 1835–1843.

McCarville, M.T., O'neal, M.E., Potter, B., Tilmon, K.J., Cullen, E.M., Mccornack, B.P., Tooker, J.F., Prischmann-Voldseth, D., 2014. One gene versus two: a regional study on the efficacy of single gene versus pyramided resistance for soybean aphid management. Journal of Economic Entomology 107, 1680–1687.

McDonald, B., Linde, C., 2002. The population genetics of plant pathogens and breeding strategies for durable resistance. Euphytica 124, 163–180.

McIntosh, R.A., Brown, G., 1997. Anticipatory breeding for resistance to rust diseases in wheat. Annual Review of Phytopathology 35, 311–326.

Metcalf, R., 1986. Foreword. In: Krysan, J.L., Miller, T.A. (Eds.), Methods for the Study of Pest Diabrotica. Springer-Verlag, New York, pp. vii-xv.

Miah, G., Rafii, M.Y., Ismail, M.R., Puteh, A.B., Rahim, H.A., Asfaliza, R., Latif, M.A., 2013. Blast resistance in rice: a review of conventional breeding to molecular approaches. Molecular Biology Reporter 40, 2369–2388.

Miles, M.R., Bonde, M.R., Nester, S.E., Berner, D.K., Frederick, R.D., Hartman, G.L., 2011. Characterizing resistance to *Phakopsora pachyrhizi* in soybean. Plant Disease 95, 577–581.

Miles, M.R., Frederick, R.D., Hartman, G.L., 2006. Evaluation of soybean germplasm for resistance to *Phakopsora pachyrhizi*. Plant Health Progress. http://dx.doi.org/10.1094/PHP-2006-0104-01-RS.

Miles, M.R., Levy, C., Morel, W., Mueller, T., Steinlage, T., Van Rij, N., Frederick, R.D., Hartman, G.L., 2007. International fungicide efficacy trials for the management of soybean rust. Plant Disease 91, 1450–1458.

Misener, G.C., Boiteau, G., McMillan, L.P., 1993. A plastic-lining trenching device for the control of Colorado potato beetle: beetle excluder. American Potato Journal 70, 903–908.

Molinari, S., Baser, N., 2010. Induction of resistance to root-knot nematodes by SAR elicitors in tomato. Crop Protection 29, 1354–1362.

Moose, S., Mumm, R., 2008. Molecular plant breeding as the foundation for 21st century crop improvement. Plant Physiology 147, 969–977.

Mundt, C., 2014. Durable resistance: a key to sustainable management of pathogens and pests. Infection, Genetics and Evolution 27, 446–455.

Munkvold, G., Hellmich, R., 1999. Genetically modified insect resistant corn: Implications for disease management. APSnet Features. http://www.apsnet.org/publications/apsnetfeatures/Pages/InsectResistantCorn.aspx [Online] (accessed October 2015).

Ocsoy, I., Paret, M.L., Ocsoy, M.A., Kunwar, S., Chen, T., You, M., Tan, W., 2013. Nanotechnology in plant disease management: DNA-directed silver nanoparticles on graphene oxide as an antibacterial against *Xanthomonas perforans*. ACS Nano 7, 8972–8980.

Oerke, E.C., 2005. Crop losses to pests. The Journal of Agricultural Science 144, 31–43.

Oliver, R.P., Ipcho, S.V.S., 2004. *Arabidopsis* pathology breathes new life into the necrotrophs-vs.-biotrophs classification of fungal pathogens. Molecular Plant Pathology 5, 347–352.

Păcurar, D.I., Thordal-Christensen, H., Păcurar, M.L., Pamfil, D., Botez, C., Bellini, C., 2011. *Agrobacterium tumefaciens*: from crown gall tumors to genetic transformation. Physiological and Molecular Plant Pathology 76, 76–81.

Palloix, A., Ayme, V., Moury, B., 2009. Durability of plant major resistance genes to pathogens depends on the genetic background, experimental evidence and consequences for breeding strategies. New Phytologist 183, 190–199.

Pardo-López, L., Soberón, M., Bravo, A., 2013. *Bacillus thuringiensis* insecticidal three-domain Cry toxins: mode of action, insect resistance and consequences for crop protection. FEMS Microbiology Reviews 37, 3–22.

Paul, C., Bowen, C.R., Bandyopadhyay, R., Tefera, H., Adeleke, R., Sikora, E., Pegues, M.D., Hartman, G.L., 2010. Registration of three soybean germplasm lines resistant to *Phakopsora pachyrhizi* (soybean rust). Journal of Plant Registrations 4, 244–248.

Pedersen, W.L., Leath, S., 1988. Pyramiding major genes for resistance to maintain residual effects. Annual Review of Phytopathology 26, 369–378.

Pernezny, K., Litz, R., 2003. Some Common Diseases of Papaya in Florida. Plant Pathology Department Document PP35. Florida Cooperative Extension Service, Institute of Food and Agricultural Sciences, University of Florida.

Petzoldt, C., Seaman, A., 2006. Climate change effects on insects and pathogens. Climate Change and Agriculture: Promoting Practical and Profitable Responses. Available from: http://umaine.edu/oxford/files/2012/01/III.2Insects.Pathogens1.pdf.

Pieterse, C.M.J., Van Wees, S.C.M., 2015. Induced disease resistance. In: Lugtenberg, B. (Ed.), Principles of Plant-Microbe Interactions: Microbes for Sustainable Agriculture. Springer International Publishing.

Pietrusińska, A., Czembor, P.C., Czembor, J.H., 2013. *Lr39* + *Pm21*: a new effective combination of resistance genes for leaf rust and powdery mildew in wheat. Czech Journal of Genetics and Plant Breeding 49, 109–115.

Pink, D., 2002. Strategies using genes for non-durable disease resistance. Euphytica 124, 227–236.

Porter, D.R., Burd, J.D., Shufran, K.A., Webster, J.A., 2000. Efficacy of pyramiding greenbug (Homoptera: Aphididae) resistance genes in wheat. Journal of Economic Entomology 93, 1315–1318.

Pradhanang, P., Ji, P., Momol, M., Olson, S., Mayfield, J., Jones, J., 2005. Application of acibenzolar-*S*-methyl enhances host resistance in tomato against *Ralstonia solanacearum*. Plant Disease 89, 989–993.

Pradzynska, A., 1982. The suitability of ultrasound for controlling stored pest. Prace Naukowe Institute Ochrony Roslin 24, 77–90.

Ramamoorthy, V., Viswanathan, R., Raguchander, T., Prakasam, V., Samiyappan, R., 2001. Induction of systemic resistance by plant growth promoting rhizobacteria in crop plants against pests and diseases. Crop Protection 20, 1–11.

Raupach, G.S., Kloepper, J.W., 1998. Mixtures of plant growth-promoting rhizobacteria enhance biological control of multiple cucumber pathogens. Phytopathology 88, 1158–1164.

Raupach, G.S., Kloepper, J.W., 2000. Biocontrol of cucumber diseases in the field by plant growth-promoting rhizobacteria with and without methyl bromide fumigation. Plant Disease 84, 1073–1075.

Romero, A.M., Kousik, C.S., Ritchie, D.F., 2001. Resistance to bacterial spot in bell pepper induced by acibenzolar-*S*-methyl. Plant Disease 85, 189–194.

Ross, A.F., 1961a. Localized acquired resistance to plant virus infection in hypersensitive hosts. Virology 14, 329–339.

Ross, A.F., 1961b. Systemic acquired resistance induced by localized virus infections in plants. Virology 14, 340–358.

Samborski, D., 1985. Wheat leaf rust. The Cereal Rusts 2, 39–59.

Sardesai, N., Nemacheck, J., Subramanyam, S., Williams, C., 2005. Identification and mapping of *H32*, a new wheat gene conferring resistance to Hessian fly. Theoretical and Applied Genetics 111, 1167–1173.

Schafer, J., Roelfs, A., 1985. Estimated relation between numbers of urediniospores of *Puccinia graminis* f. sp. *tritici* and rates of occurrence of virulence. Phytopathology 75, 749–750.

Scheuermann, K., Raimondi, J., Marschalek, R., De Andrade, A., Wickert, E., 2012. *Magnaporthe oryzae* genetic diversity and its outcomes on the search for durable resistance. In: Caliskan, M. (Ed.), Molecular Basis for Plant Genetic Diversity. InTech. Available from: http://www.intechopen.com/books/the-molecular-basis-of-plant-geneticdiversity/magnaporthe-oryzae-genetic-diversity-and-its-outcomes-on-the-search-for-durable-resistance.

Servin, A., Elmer, W., Mukherjee, A., de la Torre-Roche, R., Hamdi, H., White, J.C., Bindraban, P., Dimkpa, C., 2015. A review of the use of engineered nanomaterials to suppress plant disease and enhance crop yield. Journal of Nanoparticle Research 17, 1–21.

Schneider, R.W., Hollier, C.A., Whitam, H.K., Palm, M.E., Mckemy, J.M., Hernandez, J.R., Levy, L., Devries-Paterson, R., 2005. First report of soybean rust caused by *Phakopsora pachyrhizi* in the continental United States. Plant Disease 89, 774.

Servin, B., Martin, O.C., Mézard, M., 2004. Toward a theory of marker-assisted gene pyramiding. Genetics 168, 513–523.

Shi, A., Chen, P., Li, D., Zheng, C., Zhang, B., Hou, A., 2009. Pyramiding multiple genes for resistance to soybean mosaic virus in soybean using molecular markers. Molecular Breeding 23, 113–124.

Sillero, J., Rojas-Molina, M., Avila, C., Rubiales, D., 2012. Induction of systemic acquired resistance against rust, ascochyta blight and broomrape in faba bean by exogenous application of salicylic acid and benzothiadiazole. Crop Protection 34, 65–69.

Sticher, L., Mauch-Mani, B., Métraux, J.P., 1997. Systemic acquired resistance. Annual Review of Phytopathology 35, 235–270.

Stone, M., 2010. Virulent new strains of rust fungus endanger world wheat. Microbe 5, 423–428.

Sudisha, J., Sharathchandra, R.G., Amruthesh, K.N., Kumar, A., Shetty, H.S., 2012. Pathogenesis related proteins in plant defense response. Biological Control 12, 379–403.

Suh, J.-P., Jeung, J.-U., Noh, T.-H., Cho, Y.-C., Park, S.-H., Park, H.-S., Shin, M.-S., Kim, C.-K., Jena, K.K., 2013. Development of breeding lines with three pyramided resistance genes that confer broad-spectrum bacterial blight resistance and their molecular analysis in rice. Rice Science 6, 1–11.

Sundaram, R.M., Vishnupriya, M.R., Biradar, S.K., Laha, G.S., Reddy, G.A., Rani, N.S., Sarma, N.P., Sonti, R.V., 2008. Marker assisted introgression of bacterial blight resistance in Samba Mahsuri, an elite indica rice variety. Euphytica 160, 411–422.

Tabashnik, B.E., Brévault, T., Carrière, Y., 2013. Insect resistance to Bt crops: lessons from the first billion acres. Nature Biotechnology 31, 510–521.

Tanweer, F.A., Rafii, M.Y., Sijam, K., Rahim, H.A., Ahmed, F., Latif, M.A., 2015. Current advance methods for the identification of blast resistance genes in rice. Comptes Rendus Biologies 338.

Tar'an, B., Buchwaldt, L., Tullu, A., Banniza, S., Warkentin, T., Vandenberg, A., 2003. Using molecular markers to pyramid genes for resistance to ascochyta blight and anthracnose in lentil (*Lens culinaris* Medik). Euphytica 134, 223–230.

Tennant, P., Gonsalves, C., Ling, K., Fitch, M., Manshardt, R., Slightom, J., Gonsalves, D., 1994. Differential protection against papaya ringspot virus isolates in coat protein gene transgenic papaya and classically cross-protected papaya. Phytopathology 84, 1359–1365.

Terenius, O., Papanicolaou, A., Garbutt, J.S., Eleftherianos, I., Huvenne, H., Kanginakudru, S., Albrechtsen, M., An, C., Aymeric, J.-L., Barthel, A., 2011. RNA interference in Lepidoptera: an overview of successful and unsuccessful studies and implications for experimental design. Journal of insect physiology 57, 231–245.

Tripathi, S., Suzuki, J.Y., Ferreira, S.A., Gonsalves, D., 2008. Papaya ringspot virus–P: characteristics, pathogenicity, sequence variability and control. Molecular Plant Pathology 9, 269–280.

Tyagi, S., Mir, R.R., Kaur, H., Chhuneja, P., Ramesh, B., Balyan, H.S., Gupta, P.K., 2014. Marker-assisted pyramiding of eight QTLs/genes for seven different traits in common wheat (*Triticum aestivum* L.). Molecular Breeding 34, 167–175.

Udayashankar, A., Chandra Nayaka, S., Reddy, M., Srinivas, C., 2011. Plant growth-promoting rhizobacteria mediate induced systemic resistance in rice against bacterial leaf blight caused by *Xanthomonas oryzae* pv. *oryzae*. Biological Control 59, 114–122.

Valder, G., 1893. Experiments with pulses: soy bean (*Soja hispdia*, Moench). Agricultural Gazette of New South Wales 4, 915–916.

Vallad, G.E., Goodman, R.M., 2004. Systemic acquired resistance and induced systemic resistance in conventional agriculture. Crop Science 44, 1920–1934.

Vincent, C., Panneton, B., Fleurat-Lessard, F. (Eds.), 2001. Physical Control in Plant Protection. Springer/INRA, Berlin/Paris, 329 p.

Vincent, C., Hallman, G., Panneton, B., Fleurat-Lessard, F., 2003. Management of agricultural insects with physical control methods. Annual Review of Entomology 48, 261–281.

Vitti, A., La Monaca, E., Sofo, A., Scopa, A., Cuypers, A., Nuzzaci, M., 2014. Beneficial effects of *Trichoderma harzianum* T-22 in tomato seedlings infected by *Cucumber mosaic virus* (CMV). BioControl 1–13.

Wei, G., Kloepper, J.W., Tuzun, S., 1996. Induced systemic resistance to cucumber diseases and increased plant growth by plant growth-promoting rhizobacteria under field conditions. Phytopathology 86, 221–224.

White, R.F., 1979. Acetylsalicylic acid (aspirin) induces resistance to *Tobacco mosaic virus* in tobacco. Virology 99, 410–412.

Yedidia, I., Shoresh, M., Kerem, Z., Benhamou, N., Kapulnik, Y., Chet, I., 2003. Concomitant induction of systemic resistance to *Pseudomonas syringae* pv. *lachrymans* in cucumber by *Trichoderma asperellum* (T-203) and accumulation of phytoalexins. Applied and Environmental Microbiology 69, 7343–7353.

Yeh, S.-D., Gonsalves, D., 1984. Evaluation of induced mutants of papaya ringspot virus for control by cross protection. Phytopathology 74, 1086–1091.

Yu, W., Han, F., Gao, Z., Vega, J., Bircher, J., 2007. Construction and behavior of engineered minichromosomes in maize. Proceedings of the National Academy of Sciences United States of America 104, 8924–8929.

Zehnder, G.W., Murphy, J.F., Sikora, E.J., Kloepper, J.W., 2001. Application of rhizobacteria for induced resistance. European Journal of Plant Pathology 107, 39–50.

Ziebell, H., Carr, J.P., 2010. Cross-protection: a century of mystery. Advances in virus research 76, 211–264.

The Use of Technologies for Sufficient and Quality Animal-Food Production

4

Kevin M. Wade
Department of Animal Science, McGill University, QC, Canada

1. Introduction

Food security and availability has, perhaps, never posed as enormous a challenge as it does at this moment in time. Faced with a growing population of diverse cultures, with hugely variable economic and technological resources, the ability to provide an abundant future food supply—both plant and animal sources—is one of the greatest challenges of our generation. The need for more food is unquestionable. What is not clear is how the necessary increases from animal food sources can be achieved under the significant challenges that face the livestock industry: the provision of affordable, quality food products within an environment that respects animal well-being and the demands of consumers.

This chapter will examine three key areas of food production from animal sources:

1. The need for quality animal products in terms of a healthy diet, provision of essential nutrients, and the undeniable market for animal products as disposable income increases.
2. The use of verified scientific methods and technologies to meet this increased need through genetic selection and modification, reproductive technologies, satisfying the nutritional needs of the animal, and providing a healthy, disease-free, and respectful environment for animal raising and harvest.
3. The challenges that the use of such technologies brings within a context of limited physical resources, increased urbanization, problems of waste and distribution, consumer disconnect, and the dangers of minority groups over-influencing the needs of the majority.

2. The Need for Animal Products

Asking the question "do we have enough, good food?" seems so straightforward. However, those two simple adjectives that signify quantity and quality are dependent on so many variables and contexts that the answer is not only extremely complicated, but it is virtually impossible to answer. Even if we were to fix the term "quantity" by some objective measure like energy (e.g., calories), the term "quality" often conjures up many different images: it can have connotations of nutrition, taste, fresh versus preserved/processed, and, increasingly often, a sense of safety. Moreover, many may

Emerging Technologies for Promoting Food Security. http://dx.doi.org/10.1016/B978-1-78242-335-5.00004-4

assign such characteristics as shelf life, sustainability of production, and environmental cost to the concept of quality.

In addition, asking such a question in Montreal, New York, or Los Angeles is already likely to elicit a significant variation in the responses; regardless of their all being well-developed cities with high standards of living, the effects of lifestyle, climate, and ethnic makeup are all likely to influence the notion of what is "good food" and how much is "enough." To consider that same question in cities like Dhaka, Addis Ababa, or Djibouti adds many more dimensions to the complexity of the situation, bringing in such additional aspects as culture, land resources, politics, economics, standard of living, and even religious influences. Now consider that same question but limited to the context of animal food!

It is not the goal of this chapter to argue for or against the ethics of consuming animal products. Clearly there are groups of individuals who would have us eliminate animals completely from the human food chain. Consideration of this view is certainly required when examining the challenges of animal production (and will be addressed in Section 4); however, it is unreasonable to believe that such an occurrence will prevail on a worldwide scale. We can, and should, debate the actual *amounts* of animal-food products that are produced and consumed, but their continued contribution to the human food chain seems inevitable. There are various reasons why animal products will continue to be promoted as part of a healthy diet (Bender, 1992); standard reasons range from the quality of the protein (adequate amounts of essential amino acids) to various vitamins (e.g., B_{12}) that cannot be manufactured by the human body (unless we plan on exploring the consumption of fecal matter!), to simple ease of digestibility by the monogastric stomach.

In addressing the question of quantity first, Health Canada recommendations for daily caloric intake can be simplified to 1500, 2000, and 2500 for children (1–15 years), older females, and older males, respectively (Canada's Food Guide, 2015). A simplified extrapolation to a world-population basis (7.2 billion), assuming 25% children and 37.5% each for older males and females, gives an approximate value of global caloric needs on a daily and yearly basis.

$$[(1500 * 0.25) + (2250 * 0.75)] * 7\,200\,000\,000 = 1.485 \times 10^{13} \text{ Calories per day or } 5.42025 \times 10^{15} \text{ Calories per year.} \quad (1)$$

While there is some evidence that the use of our world crops for human consumption alone (i.e., bypassing animals) could provide not only the requisite numbers of calories and protein amounts for the current population of 7.2 billion, but even for future growth (Conte, 2015), even here there is some question as to the requisite amount of fat that is required for daily diets. In addition, there are sound nutritional reasons (in terms of protein, minerals, and vitamins) why dairy and meat products (or some alternatives) are important components of our diets, especially with regard to young children and pregnant/lactating females (Bender, 1992; FAO, 2011b). Health Canada acknowledges this in their recommended servings, even though every country's recommendations could be argued to be based on its access to specific food sources (Table 1).

Table 1 **Recommended Number of Food-Guide Servings per Day**

	Children			Teens		Adults			
	2–3	4–8	9–13	14–18 years		19–50 years		51+ years	
	Girls and Boys			Female	Male	Female	Male	Female	Male
Vegetables and fruit	4	5	6	7	8	7–8	8–10	7	7
Grain products	3	4	6	6	7	6–7	8	6	7
Milk and alternatives	2	2	3–4	3–4	3–4	2	2	3	3
Meat and alternatives	1	1	1–2	2	3	2	3	2	3

Source: Eating Well with Canada's Food Guide. Health Canada, 2011. Adapted and reproduced with permission from the Minister of Health (2015).

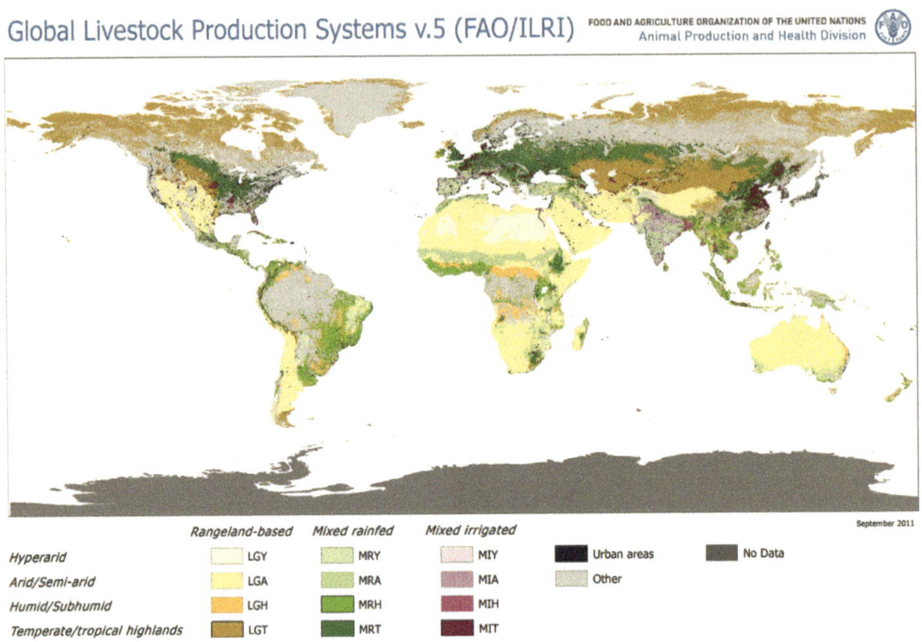

Figure 1 Global classification of land.
Reproduced with permission from FAO GeoNetwork: Global Livestock Production Systems v.5 2011, http://www.fao.org/ag/againfo/resources/en/glw/GLW_prod-sys.html.

Regardless of the mathematical formula (1) or actual extrapolated numbers, it is clear that political and distribution problems complicate such an argument. The solution of simply giving food to needy areas has created many additional challenges related to importation challenges, equitable distribution, competition with local developing economies, and a culture of dependency (FAO, 2012c). Production and distribution is further exacerbated by the fact that arable land is not evenly distributed across the globe (Figure 1). In fact, of the estimated 13 billion ha of land in the world, less than 11% is made up of arable land. More strikingly, perhaps, is the fact that more than two-thirds of our agricultural land is classified as "permanent meadows and pastures"—land that is not suitable for large-scale crop production (Table 2, Figure 2). If this land is to be exploited for the necessary large-scale production of food, it is ruminant livestock that offer the best avenue for transforming these resources into food for human consumption (Eisler et al., 2014) where even low-input systems present an opportunity for the production of quality food.

According to FAO, WFP, and IFAD (2012), economic and agricultural growth should be "nutrition sensitive":

> *Growth needs to result in better nutritional outcomes through enhanced opportunities for the poor to diversify their diets; improved access to safe drinking water and sanitation; improved access to health services; better consumer awareness regarding adequate nutrition and child care practices; and targeted distribution of supplements in situations of acute micro-nutrient deficiencies. Good nutrition, in turn, is key to sustainable economic growth.*

Table 2 **A Breakdown of the Earth's Land Surface**

Land	Area (ha)
Land area	13,009,101,790
Agricultural area	4,922,206,560
Arable land and permanent crops	1,562,547,970
Arable land	1,395,894,500
Permanent crops	163,893,470
Permanent meadows and pastures	3,359,658,590
Forest area	4,021,910,770
Other land	4,074,444,530
Inland water	457,437,600

Data from FAOSTAT, http://faostat.fao.org/site/377/DesktopDefault.aspx?PageID=377#ancor (accessed 15.02.15.).

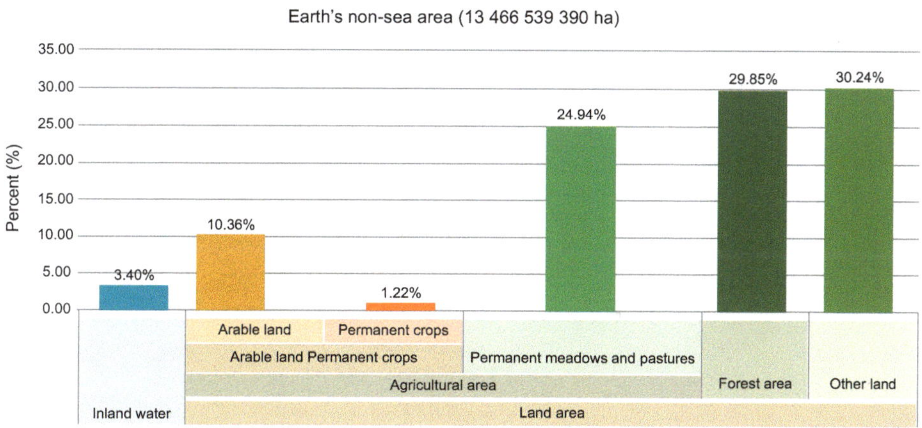

Figure 2 Non-sea area of the earth.
Data from FAOSTAT, http://faostat.fao.org/site/377/DesktopDefault.aspx?PageID=377#ancor (accessed 15.02.15.).

Ironically, perhaps, the strongest argument that demand for animal products will increase is only *indirectly* related to nutritional and delivery claims; all the evidence suggests that as societies become wealthier, the amount of salary spent on meat products increases (Table 3, Figure 3). With many large regions of the world poised to make the transition from developing to developed, this poses an enormous challenge for the future production and delivery of sufficient animal-food products. An increase in China's gross domestic product (GDP) alone over the next 10 years would account for significant pressure on the current system of animal production under such a scenario (Millar, 2010). Regardless of the pros and cons of animal meat and dairy products, if viewed as luxury items, they will continue to occupy a strong market advantage.

In addition to enormous growth of populations in developing countries, the increasing discrepancy between rural and urban populations will continue to play a huge role in how food is produced, processed, distributed, and consumed. We already have a majority of

Table 3 Per Capita Income and Population Will Grow Faster in Developing Countries than in Developed Countries

	Per capita Income		Population	
	2012 Level (2005 dollars)	2013–2022 Growth Rate (%)	2012 Level (million)	2013–2022 Growth Rate (%)
World	7745	2.4	6913	1.0
High-income countries	37,594	1.7	927	0.4
All developing countries	2906	4.4	5582	1.2
Latin America	5959	3.1	602	1.0
Asia (excluding Japan)	2572	5.8	3618	0.9
Middle East	6743	2.5	298	1.4
Africa	1254	2.6	1052	2.2

Source: United States Department of Agriculture, Economic Research Service, International Macroeconomic Data Set.

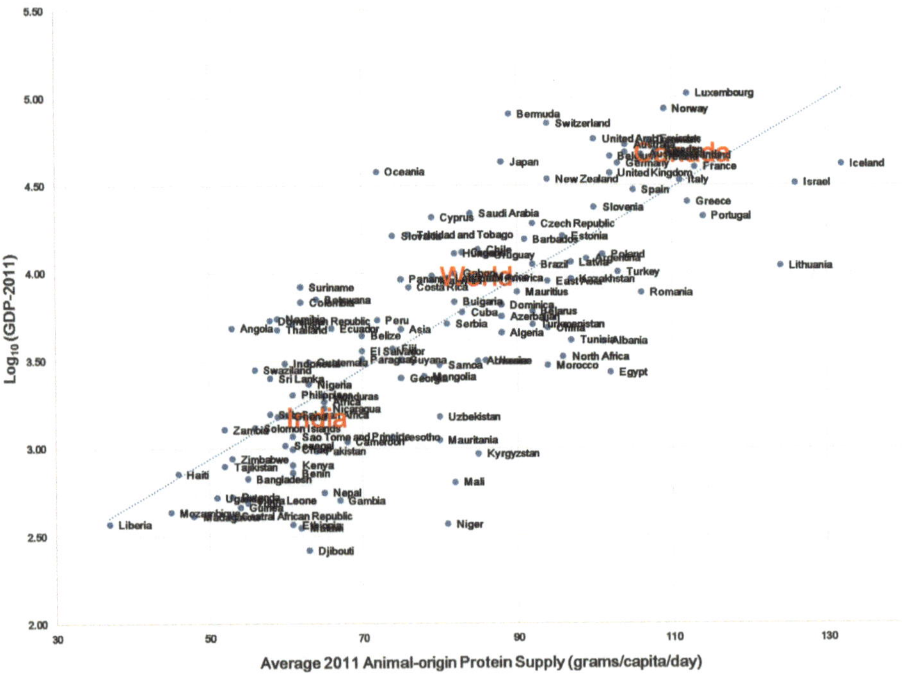

Figure 3 Countries showing relationship between amount of protein supplied from animal origins versus Log(10) GDP (2011 Figures) highlighting values for the World, India, and Canada.
Data from USDA.

the world's population living in urban areas—54%. This compares to 30% in 1950 and is estimated to grow to 66% by 2050 (United Nations, 2014). While a 2014 average of 54% may not (or may!) strike one as alarming, it is important to note the variation, based on regions of the world: North America (82%), Latin America (80%), and Europe (73%). These regions also include some of the richer economies, which implies that the demand for meat products is higher as well. In contrast, Africa and Asia are still primarily rural (60% and 52%, respectively), but these percentages are changing more quickly than in the developed world and are expected to drop considerably by 2050 (44% and 36% rural, respectively). These regions also contain the largest population concentrations in the world (China and India). In fact, according to World Urbanization Prospects, Revision 2014 (United Nations, 2014):

> *Just a few countries are home to half of the world's urban population. China has the largest urban population (758 million), followed by India (410 million). These two countries account for 30 percent of the world's urban population and, with another five countries, the United States of America (263 million), Brazil (173 million), Indonesia (134 million), Japan (118 million) and the Russian Federation (105 million), account for more than half of the world's urban population.*

With an estimated two-thirds of the 2050 population predicted to be living in urban settings, this poses another challenge for the provision of meat products to those societies. While there is certainly poverty in large urban areas, the average urban income tends to be higher than its rural counterpart (FAO, 2011b). Urban dwellers are generally pure consumers of products, whereas rural communities are more likely to produce and consume. Due to the wealthier urban lifestyle, the demand for meat products tends to be higher, and the large-scale distribution schemes allow easy access to a variety of these products in efficient formats.

With all these factors in mind, it is clear that not only the overall amount of food needs to increase considerably by 2050 and beyond, but also the amount of food from animal sources. Many authors forecast that we will need to double our current amounts of food (Simmons, 2010, 2011; FAO, 2012a; NABC, 1999) in order to meet the global needs of some 9 to 10 billion inhabitants. With regard to meat and dairy products specifically, Alexandratos and Bruinsma (2012) estimate that the amount of meat needed to feed the 2050 population will be 455 million tons (up from 2007 figures of 258 million tons). In fact, FAO (2011b) estimates an increase of 173% in overall meat products by 2050 (compared to 2010), ranging from 137% for pig meat to 225% for poultry meat.

This is a daunting challenge in any context—plant or animal—and requires an in-depth analysis of our resources and our means to exploit them to the maximum, within the ever-increasingly complicated contexts of politics, culture, and consumer attitudes.

3. Improved Animal Production Through Technology

There have already been staggering increases in animal production systems over time. Taking milk production as an example, the average milk production per cow in the United States in the early 1940s was just over 2000 kg, and had risen to just over

9000kg by 2007—a greater than fourfold increase in the producing ability of a dairy cow (Capper et al., 2009). Similar trends exist for Canada, with consistent increases of 50kg per cow per year over the last decade (Canadian Dairy Information Centre, 2014 data). This improvement is not restricted to dairy products; during the period of 1925–1995 in the United States, the number of eggs produced by a hen in a year, carcass weight in beef cattle, and market live weight in sheep have all more than doubled; market live weight in pigs has almost tripled; and age to market weight in broilers has dropped from 15 to 7 weeks (Field and Taylor, 2010). In terms of our ability to produce more from less, these are phenomenal rates of improvement over time.

There is an old adage that "half of the improvement comes from breeding, and half comes from feeding." While genetics and nutrition have played very significant roles in the improvement of livestock production, there are other domains involved. In fact, it is probably more accurate to assign historical percentages of 35% to genetic selection, 35% to improved animal nutrition, 15% to reproductive technologies, and 15% to everything else (housing, health, comfort, etc.). With the wave of research in biotechnology and a better understanding of the underlying physiology of the various livestock involved, there is still the potential for greater improvement in the kinds of livestock we produce and in the ways we manage them for production.

3.1 Genetic Improvement

Traditional selection for traits of interest really began as soon as man understood Darwin's concepts of natural selection and survival of the fittest. It meant that by defining the environment in which the animal was to "succeed," the best animals could be selected to produce future generations, thus adapting to the management system. This sense of artificial selection has been the cornerstone for improvement in traits of interest since the mid-1800s. Mendel's groundbreaking discoveries into the functionality of genes and their basic behavior allowed for selection theory to develop and for tangible predictions to be produced over time. In essence, breeders were attempting to improve characteristics in their livestock that provided a better product for sale (thereby increasing profit). Of course, this depended on how much the improvement of the "trait" was based on a quantifiable genetic basis, and how much was solely due to the management conditions (environment) under which the animal was raised (i.e., heritability). As breeding strategies became more ambitious and selection was attempted on multiple traits (e.g., heavier weaning weight, better feed efficiency, larger market weight, better quality carcass, etc.), determining just how much was determined by the genetic makeup and how much was due to management became a real challenge. In addition, even when there was a solid genetic basis for two different traits, negative genetic correlations often stymied overall progress in both. Selection index theory provided a manner by which weighted emphasis could be placed on the traits, based on their economic importance, thereby identifying animals that provided the best opportunity for *overall* improvement of the traits in question—even though the best overall was not necessarily the best for each trait. There has been significant genetic improvement in most livestock traits for which there is a genetic basis and for which the selection strategy has been consistent over time.

As an example, multiple selection in dairy-cattle traits broadly concentrates on milk production (51%), length of life in the dairy herd (34%), and health and fertility traits (15%) (Canadian Dairy Network, 2013).

Figure 4 shows the progress in these three areas in the Canadian dairy population, as well as the overall index, since 1991 when the Lifetime Profit Index was introduced (Canadian Dairy Network, 2013). While any index is only as good as the relevance of the included traits and the accuracy of the economic weights that are attached to them, it should be noted that there has been a considerable increase in the Lifetime Profit Index over the last 25 years. The degree of improvement reflects the genetic component of the groups of traits being selected, and an obvious example of the challenge with genetic improvement is seen in the relatively small progress for health and fertility. Introduced in 1998 as an area of importance to dairy producers, its genetic progress is a function of the low heritabilities associated with some of its components (somatic cell score, udder depth, milking speed, daughter fertility, and lactation persistency). This essentially means that good health and fertility in an animal have more to do with effective management, estrus detection, sanitation, and housing conditions than with aspects of an animal's genome that can be readily selected.

This move to a selection for more functional traits (rather than sole selection on the traits of interest to consumers) represents a real technological advance, as we understand the importance of an animal's overall condition and its ability adapt to a more intense environment. This continues to have an influence on the cost of production (through decreased health costs) and a shorter generation interval that comes with a better understanding of how to manage fertility. Both of these aspects result in the production of meat, dairy, and egg products in a more cost-effective manner that can be

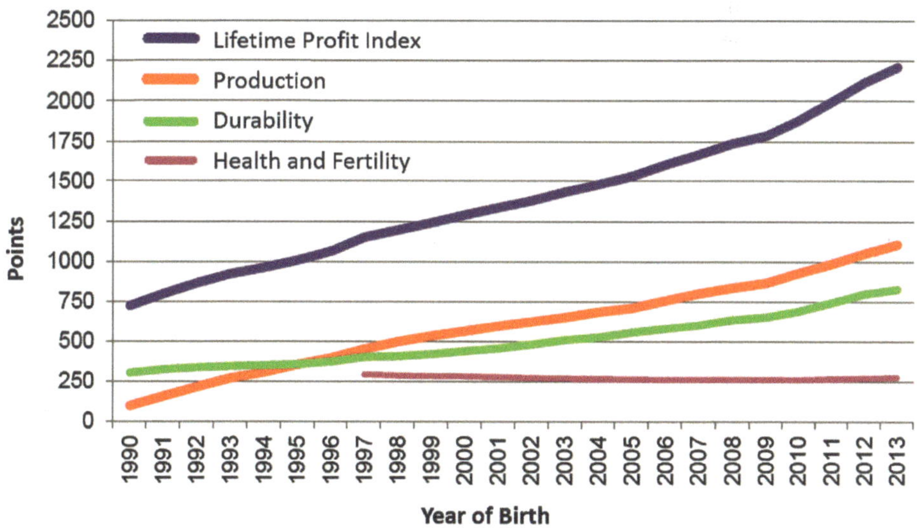

Figure 4 Points for Lifetime Profit Index, production, durability, and health and fertility in the Canadian dairy population (1991–2013).
Adapted from Canadian Dairy Network et al. (2013).

Table 4 **First Sequenced Genomes (and Details) for Selected Animal Species**

Organism/Name	Size (Mb)	GC%	Genes	Proteins	Release Date
Chicken (*Gallus gallus*)	1046.93	41.93	21,158	32,182	29–02–2004
Rabbit (*Oryctolagus cuniculus*)	2737.46	44.05	28,903	37,564	13–05–2005
Horse (*Equus caballus*)	2474.93	41.65	25,390	32,352	24–01–2007
Pig (*Sus scrofa*)	2808.53	42.45	34,633	38,370	11–07–2008
Cattle (*Bos taurus*)	2670.14	41.89	36,107	52,038	24–04–2009
Turkey (*Meleagris gallopavo*)	1128.34	41.72	20,903	26,423	07–09–2010
Sheep (*Ovis aries*)	2619.05	42.00	48,452	44,976	21–09–2012

Adapted from The National Center for Biotechnology Information.

passed on through the chain to the consumer. It ultimately means that the same output can be produced in a shorter time period and with lower costs of production.

The second big technological advance relates to the availability of genome sequences for many of the major livestock species (Table 4). This has allowed for a more complete understanding of the effects of genes on traits of interest, ranging from quantitative trait locus studies (e.g., Goddard and Hayes, 2009) to the latest approaches in genomic selection (e.g., de Roos et al., 2011), and a corresponding increase in earlier and more accurate prediction of superior animals for selection programs (Hayes et al., 2009). In essence, we are approaching an ability to pinpoint regions of the genome that have demonstrated control over areas of economic utility, and we can associate those regions with young untested animals. One might argue that we are closer than ever to being able to use the information that Mendel derived (regarding single genes and their mechanisms of effect) in a more direct manner (as opposed to solely from phenotypic selection). As with every good tool, however, there are challenges with its use, and some authors have already pointed out the potential risks of bias in selection, increased rates of inbreeding, and the danger of distancing the scientist from the producer in terms of novel data collection and an inclusion of the farm's microenvironment (Dürr and Philipsson, 2012; Dechow, 2014). Nevertheless, a greater understanding of a species' genome represents a key technological advance as we begin to target the improvement of specific genetic makeups in our livestock, dependent on individual production systems and environments. The possibility exists of developing lines of the same species for explicitly defined products, depending on the market and the needs of a specific society.

Advances in cross-breeding have also been put to good use: the development of new breeds that combine characteristics—otherwise difficult to select in one single breed, due to negative genetic correlations—is allowing for the production of quality animal products in areas where there was a need before. An excellent example of these technologies is the development of the Barangus—a cross that is 3/8 Brahman (*Bos indicus*) and 5/8 Angus (*Bos taurus*) (Briggs and Briggs, 1980). They combine the good carcass composition of the Angus with the heat tolerance, insect resistance, and general hardiness of the Brahman (Figure 5). Cross-breeding techniques can be used for terminal meat products, combination of meat and maternal traits, and, as in the case of the Brangus, new breeds that provide a more economical product in a potentially unfavorable environment.

Figure 5 A Brangus cow with calf by Scott Bauer, USDA ARS.
Licensed under Public Domain via Wikimedia Commons; http://commons.wikimedia.org/wiki/File:Cow_with_calf.jpg#mediaviewer/File:Cow_with_calf.jpg.

3.2 Nutritional Needs

Selection of animals for targeted purposes is only a starting point; they must subsequently be raised to produce the product of interest—themselves (in the case of most meat production systems) or their products (dairy and eggs). Great strides have been made in the understanding of ruminant nutrition and how improved feed efficiency can affect many management aspects, including animal health (Eastridge, 2006) and welfare (FAO, 2012b). The National Research Council (U.S.) has developed a series of publications, written by recognized experts on nutrition subcommittees under the auspices of the Board of Agriculture and Natural Resources, regarding the nutrient requirements of various livestock species: Sheep (1975); Rabbits (1977); Goats (1981); Poultry (1994); Beef Cattle (2000); Dairy Cattle (2001); Horses (2007); Fish and Shrimp (2011); and Swine (2012). These publications deal with each species' need for the six basic nutrient categories (water, carbohydrates, fats, proteins, minerals, and vitamins) in terms of unique production systems, different ages and growth stages, as well as their effect on the environment. In addition, commonly available feeds are assessed with regard to their provision of the required levels of nutrients, their energy for metabolic functions, and their effect on appetite (dry-matter intake). The degree of detail involved in these publications allows for the formulation of diets based on size, level of production, and even activity, providing general energy and protein requirements as well as specific mineral and vitamin quantities (see Figure 6 as an example of such detail). The fact that one can provide a ration to such a degree of detail (in conjunction with our ever-increasing knowledge of monogastric, avian, and ruminant digestive systems) allows for (1) the most appropriate diet for the intended animal products; (2) the most profitable formulation, given available ingredients; and (3) a minimum impact on the environment in terms of nutrient waste through excretion.

Since many of the meat/dairy-producing species are ruminants (and their role in the utilization of borderline arable lands has already been argued), a sound understanding of the complex interactions among plants, animals, and humans allows for a more

	Holstein = 680 kg Bwt, Mature Bwt = 680 kg, BCS = 3.0, 65 mos age, Milk fat = 3.5%, milk true protein = 3.0%, lactose = 4.8%, Default environmental conditions				Jersey = 454 kg Bwt, mature Bwt = 454 kg, BCS = 3.0, 65 mos age milk fat = 4.2%, milk true protein = 3.6%, lactose = 4.8% Default environmental conditions					
Days in milk	90	90	90	90	90	90	90	50	120	90
Dry matter intake input	Model predicted	Model predicted	Model predicted	Model predicted	Model predicted	Model predicted[h]	Model predicted	Model predicted[h]	Model predicted[h]	Model predicted + 5%[h]
Milk production (kg)	25	35	45	54.4	25	35	40	35	35	35
Milk production (lbs)	55	77	99	120	55	77	88	77	77	77
Dry matter intake (kg)	20.3	23.6	26.9	30	18	21.7	23.5	19.8	22.2	22.7
Dry matter intake (lbs)	44.7	51.9	59.2	66	39.6	47.7	51.7	43.6	48.8	49.9
Daily wt change (kg)	0.5	0.3	0.1	−0.2	0	−0.2	−0.5	−0.7	−0.1	0
Days to gain one condition score	221	316	1166		3777					4247
Days to lose one condition score				544		241	110	80	532	
Energy[a]										
NE_L (Mcal /day)	27.9	34.8	41.8	48.3	27.7	35.6	39.5	35.6	35.6	35.6
NE_L (Mcal/ kg)	1.37	1.47	1.55	1.61	1.54	1.64	1.68	1.8 a	1.6	1.57
NE_L (Mcal/ lb)	0.62	0.67	0.7	0.73	0.7	0.74	0.76	0.82	0.73	0.71
Protein										
Metabolizable protein (g/d)	1862	2407	2954	3476	1991	2639	2965	2579	2656	2672
Diet % MP	9.2	10.2	11	11.6	11.1	12.2	12.6	13	12	11.8
Rumen degradable protein (g/d)	1937	2298	2636	2947	1747	2125	2288	1971	2167	2206
Diet % RDP	9.5	9.7	9.8	9.8	9.7	9.8	9.7	10	9.8	9.7
Rumen undegradable protein (g/d)	933	1291	1677	2089	1151	1632	1865	1670	1619	1611
Diet % RUP	4.6	5.5	6.2	6.9	6.4	7.5	7.9	8.4	7.3	7.1
% RDP + % RUP (crude protein)[b]	14.1	15.2	16.0	16.7	16.1	17.3	17.6	18.4	17.1	16.8
Fiber and carbohydrate[c]										
NDF, min %	25–33	25–33	25–33	25–33	25–33	25–33	25–33	25–33	25–33	25–33
ADF, min %	17–21	17–21	17–21	17–21	17–21	17–21	17–21	17–21	17–21	17–21
NFC, max %	36–44	36–44	36–44	36–44	36–44	36–44	36–44	36–44	36–44	36–44
Minerals										
Absorbable calcium (g/day)	52.1	65	76.5	88	50.7	65.2	72.4	65.2	65.2	65.2
Dietary Ca %	0.62	0.61	0.67	0.6	0.57	0.57	0.63	0.66	0.54	0.53
Absorbable phosphorus (g/day)	44.2	56.5	68.8	80.3	41.4	54.1	60.4	52.2	54.6	55.1
Dietary P %	0.32	0.35	0.36	0.38	0.33	0.37	0.36	0.44	0.35	0.34
Mg[d] %	0.18	0.19	0.2	0.21	0.18	0.19	0.2	0.21	0.19	0.19
Cl %	0.24	0.26	0.28	0.29	0.24	0.26	0.27	0.28	0.25	0.25
K[e] %	1	1.04	1.06	1.07	1.02	1.03	1.04	1.07	1.03	1.02
Na %	0.22	0.23	0.22	0.22	0.2	0.2	0.2	0.22	0.2	0.19
S %	0.2	0.2	0.2	0.2	0.2	0.2	0.2	0.2	0.2	0.2
Co mg/kg	0.11	0.11	0.11	0.11	0.11	0.11	0.11	0.11	0.11	0.11
Cu mg/kg[f]	11	11	11	11	10	10	10	11	10	9
I mg/kg[g]	0.6	0.5	0.44	0.4	0.44	0.4	0.34	0.4	0.36	0.35
Fe mg/kg	12.3	15	17	18	14	16	17	18	16	15
Mn mg/kg	14	14	13	13	12	12	12	13	12	12
Se mg/kg	0.3	0.3	0.3	0.3	0.3	0.3	0.3	0.3	0.3	0.3
Zn mg/kg	43	48	52	55	45	49	51	54	48	47
Vitamin A (IU/day)	75000	75000	75000	75000	49500	49500	49500	49500	49500	49500
Vitamin D (IU/day)	21000	21000	21000	21000	13500	13500	13500	13500	13500	13500
Vitamin E (IU/day)	545	545	545	545	360	360	360	360	360	360
Vitamin A (IU/kg)	3685	3169	2780	2500	2772	2300	2123	2520	2247	2198
Vitamin D (IU/kg)	1004	864	758	680	755	627	579	687	613	600
Vitamin E (IU/kg)	27	23	20	18	20	17	16	18	16	16
Sample diets used in model to generate tables. Ingredients listed as kg /day DM.										
Corn silage, normal	8.48	8.21	5.61	12.02	8.96	7.77	7.39	7.1	7.96	8.15
Soybean meal, solv. 48% CP	1.01	1.62	1.41	2.39	2.16	2.78	1.67	2.54	2.85	2.91
Legume forage silage, mid-maturity	3.85	4.57			2.67	3.1		2.83	3.18	3.25
Corn grain, steam flaked	1.8	4.33	7.08	6.35	2.6	4.91	5.88	4.48	5.03	5.15
Calcium carbonate	0.04	0.07	0.09	0.02	0.06	0.04	0.03	0.04	0.04	0.04
Monosodium phosphate (1 H2O)	0.02	0.02	0.04	0.06	0.01	0.01	0.03	0.01	0.01	0.01
Soybean meal, expellers							1.16			
Legume forage hay, immature			6.16	5.42			4.59			
Sodium chloride	0.12	0.011	0.12	0.14	0.1	0.1	0.12	0.09	0.1	0.1
Grass hay, c-3, mid-mat	4.47	3.21	0.98	0.93	0.85	0.95	0.97			
Vitamin and mineral premix	0.54	0.49	0.51	0.49	0.5	0.5	0.5	0.45	0.51	0.52
Bermudagrass hay, coastal			0.87							
Cottonseed, whole with lint			2.53	2.24		1.02	1.64	0.94	1.05	1.07
Tallow				0.29		0.24	0.21	0.22	0.24	0.25
Calcium soaps of fatty acids				0.29		0.18	0.21	0.17	0.19	0.19
Blood meal, ring dried			0.23	0.31		0.11	0.1	0.1	0.11	0.12
Sorghum, sudan type, silage			2.26							
Sample diet evaluation										
NE_L (Mcal/kg)	1.49	1.55	1.57	1.58	1.54	1.59	1.57	1.62	1.58	1.57
Undiscounted TDN %	65	69	71	74	69	73	75	73	73	73

[a]Recommended energy content of early lactation rations must be limited to prevent rumen acidosis. Cow must therefore be expected to utilize body reserves to meet energy needs at highest levels of milk production. See fiber and NFC restrictions.

[b]Equivalent to crude protein requirement only if RDP and RUP are perfectly balanced.

[c]These are the minimum fiber (or maximum NFC) concentrations needed to maintain rumen health and milk fat test (see Chapter 4).

[d]Assumes that active transport of magnesium across the rumen wall is intact. High dietary potassium and excess non-protein nitrogen often interfere with Mg absorption. Under these conditions, dietary Mg should be increased to 0.3%–0.35% (see Chapter 6).

[e]Heat stress may increase the need for potassium (see Chapter 6).

[f]High dietary molybdenum, sulfur, and iron can interfere with copper absorption increasing the requirement (see Chapter 6).

[g]Diets high in goitrogenic substances increase the iodine requirement (see Chapter 6).

[h]Diet composition is the same in all four cases of the Jersey cow producing 35 kg milk. Amount of dry matter consumed varies with days in milk and the use of predicted vs. actual dry matter intake in the model.

Figure 6 An example of the degree of detail in nutrient requirements of dairy cattle. Adapted from Nutrient requirements of dairy cattle.

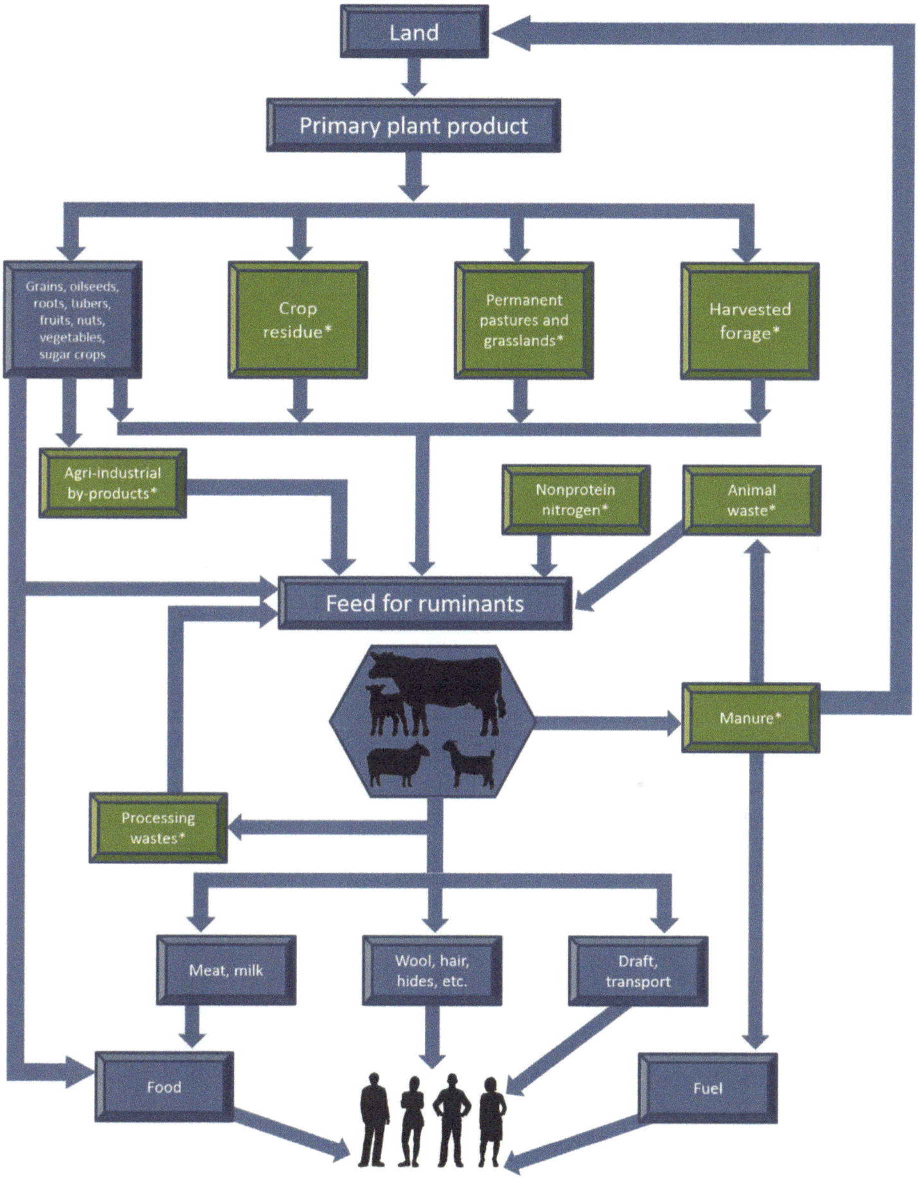

Figure 7 A graphic illustration of the land–plant–ruminant–animal–human relationship. While products in green (*) are not normally consumed by humans, ruminants can certainly convert many of them into products that are useful for humans.
Adapted from Field and Taylor (2010).

efficient use of all raw materials for human needs, a more efficient use of agri-industrial by-products, and less waste in the overall system (Figure 7). These intricate relationships are even more important when dealing with developing countries and the relatively smaller numbers of livestock.

A detailed knowledge of the nutritional needs of animals has also led to research into the modification of end products: such areas as muscle:fat ratio in meat, intramuscular fat (marbling), modification of milk fatty acids to increase the proportion of more desirable "healthy" fats from a consumer point of view, and modification of milk proteins for the industrial cheese-making industry (Bender, 1992; Lock and Bauman, 2004; Jenkins and McGuire, 2006; DePeters and Cant, 1992). Some of these goals have been achieved through optimum choice of input feed ingredients, or through bypassing the rumen (thus avoiding microbial breakdown), or by using probiotics and prebiotics to influence the rumen microflora (Benchaar et al., 2006; DiLorenzo, 2011).

Finally, our understanding of the needs of various livestock systems, and their intensive implementation in areas of the world where forages and feed sources are not naturally available, has led to the development of modified strains of plants (or even new plants in some cases) that can adapt to harsher climatic conditions, thus reducing the costs of transport and allowing production to occur in areas where it might otherwise be prohibitive (FAO, 2008).

3.3 Reproductive Physiology

While perhaps not accountable for the largest increases in livestock productivity over time (compared to genetics and nutrition), the more recent advances in physiology have allowed for great strides in the management of fertility and reproduction. There is no doubt that the development of techniques for the distribution and delivery of frozen semen through artificial insemination has allowed for enormous gains in genetic improvement (Foote, 2002). The identification of superior sires for the improvement of specific traits of importance is insignificant in the absence of a method to exploit that superiority across a breed. While not only allowing superior genetics to be distributed-virtually without limit-this technology also allows males who may have reproductive problems or low sperm counts—factors that would cause serious limitations to their promulgation in natural breeding—to be used effectively. As we look ahead to Section 4.3 and the challenges associated with some technologies, it is interesting to consider how accepted this specific technology has become with producers and consumers alike. It has, in fact, become the routine method for impregnating females with a view to breed improvement. The equivalent technology on the female side—in vitro fertilization and/or embryo transfer—also represents significant gains. While the number of eggs that can be recovered from a female (even after super-ovulatory techniques) is small, compared to sperm collection on the male side, newer technologies like harvesting of follicles from prepubertal females and in vitro embryo production have the potential to double rates of genetic gain and reduce generation intervals, especially when used in conjunction with genomic selection (Schaeffer, 2006). In addition, our increasing ability to produce sexed sperm for use with traditional AI or in vitro embryo production offers a much greater efficiency of livestock production in cases where males (meat production) or females (dairy production) may be desired almost exclusively (Garner and Seidel, 2008; Seidel, 2007; de Vries et al., 2008).

3.4 Health, Management, and Welfare

Finally, while not necessarily classified as technologies, our increasing advances in the health and comfort needs of animals make a significant contribution to their longevity and, therefore, overall profitability; an animal that is healthy will not only save veterinarian costs but will also be less prone to decreased production performance. By the same token, an animal that is kept comfortable in terms of temperature, humidity, and bedding can devote its intake energy to growth, reproduction, lactation, etc., as opposed to expending energy to, for example, keep warm. More and more, an understanding of an animal's behavioral needs (in terms of exercise, social activities, and access to "natural behavior") will usually lead to animals that perform better and stay longer in a herd (Canadian National Farm Animal Care Council).

4. The Challenges Associated with Using Improved Technologies

It is clear that not only do we need to continue to provide animal products to an ever growing world population, but we also have the technologies to both increase the amounts of products on a per unit basis and to do so more quickly than ever. Such progress is not without its challenges! These come in the form of economical, health (human and animal), ethical, and environmental challenges. In addition, perhaps the major challenge to taking advantage of increased animal production relates to industry's need to respond to minority groups who advocate avoiding products that result from these technologies. This communication with consumers needs to be carried out in a positive and proactive manner so as to avoid acrimonious debates that result in neither side listening to the other.

4.1 Economic Challenges

Mention is often made of the cost of a kg of animal protein versus the equivalent in plant protein. This is, however, complicated by such aspects as the inherent quality, their utilization of by-products, and land that is often not suitable for large-scale crop production, and the undeniable market (luxury or otherwise) for meat products, regardless of their cost. The argument that meat products are more expensive than plant products is further exacerbated by a largely urban population that is prepared to pay a significantly higher unit price for multiple preprocessed portions (e.g., boneless, skinless, marinated, etc.) rather than a purchase of the bulk item itself. It should be noted that this practice (and the lifestyle of many developed countries) leads to a significant amount of waste (Parfitt et al., 2008). It has been estimated that Africa produces 222 million tons of food (animal and plant) annually and that 230 million tons are discarded by the developed nations in terms of waste each year (FAO, 2011a). Such occurrences should be seen more as an indictment of certain societies' affluence rather than a criticism of the animal production systems that meet an accepted market need.

4.2 Health Challenges

While cost enters into the equation for organic foods in certain niches of developed countries (where a premium can often be garnered), the issue has also been raised with regard to the health of humans (and animals). Again, this is a complex issue that evokes much discussion among those who make the argument that organic products are better than those produced with fewer regulations. While there are many compounding factors that may seem to improve the quality (in a subjective sense); local versus imported, labeling, and even the environment in which such products are sold, may convince some consumers of the superior quality of the product. In fact, a review by Smith-Spangler et al. (2012) concluded that there was no strong evidence of superiority from a nutritive point of view. In addition, large-scale organic farming is unlikely to meet the same production levels as so-called conventional farming—most studies put the difference at around 80% (Halweil, 2006), although there is significant variation across different crops (Seufert et al., 2012). Given the lack of evidence of nutritional differences, this is hardly a strong argument toward our goal of doubling food production by 2050. It should, however, be stated that many of the publications that look at organic farming as a serious alternative make mention of the potential risks of pesticides and herbicides on human health (Halweil, 2006), and that a common ground approach may be tenable, whereby organic principles are followed as a matter of course rather than strict regulation. That said, there are strict regulations in most parts of the world regarding the safety of the chemicals being used and the withdrawal/waiting periods before recommended consumption. Again, this area is complicated—ironically, because of our advances in the technology of detection. The ability to detect most substances to the level of parts per billion is now standard; in fact, parts per trillion is almost routine! Admittedly, any "deadly" substances should not be present at all, and there may be cases where substances may accumulate in the body, thereby making consideration of the initial amount pertinent. That said, these are substances that have been used, often for several generations, without any conclusive evidence of their long-term effect on human (or animal) health. In addition, some societies' efforts to enact "zero-tolerance" legislation for certain substances begs the question as to what constitutes "zero." California's Proposition 37 (2012) proposed a zero-tolerance level for the adventitious presence of genetically engineered ingredients after July 1, 2019 (Carter et al., 2012). While it did not pass, one wonders how it could have been implemented or regulated in a reasonable manner. What was impossible to detect 20 years ago (therefore, officially "not there") may now be possible to detect at the level of "parts per trillion;" to put this in perspective, parts per trillion is equivalent to one drop in 50 Olympic-size swimming pools (each 50,000 m^2)! At some point, it seems that there needs to be a definition of what is a reasonable tolerance of certain substances (it may even be "zero" in some rare cases), rather than an abuse of our advances in detection technology.

4.3 Ethical Challenges

One of the biggest challenges to increasing the amount of food (both plant and animal) through technology is the debate over genetically modified organisms (GMOs). Although the most popular definition of a GMO is "any organism whose genetic material has been altered using genetic engineering techniques" (Wikipedia, 2015), there is much confusion and ignorance over the topic. As long ago as 1994, Hallman and

Metcalfe carried out a survey of New Jersey residents on the public perceptions of agricultural biotechnology. They found that while 61% of those surveyed approved of genetic engineering techniques to produce "new plants," only 28% approved of the equivalent technologies for producing new animals, and almost 50% of those in the former category were not in the latter. While more than 20 years ago, it would appear that these attitudes still prevail (Hoban, 1997; Moon and Balasubramanian, 2001; Mitchel, 2003), and the discussion has even made its way into the whole controversy of labeling (Teisl et al., 2003). Despite its stated intention of providing consumers with information and choice (as distinct from discrimination against products or legal accountability), many studies show that it does not achieve its stated goals (Carter and Gruère, 2003; Gruère, and Rao, 2007). Interestingly, in the context of providing food for a growing population, there is also some evidence to suggest that this debate is largely one of the developed countries, with wider general acceptance of genetically modified food in the developing world (Curtis et al., 2004).

The general ignorance by consumers about what is involved in GMOs (Figure 8) still prevails today, with articles on the subject introducing aspects of pesticide usage, artificial growth hormones, and even pasteurization (Teisl et al., 2003). Such ignorance is sadly compounded when certain scientific studies are carried out in a biased manner, and make improper conclusions. One of the more infamous such instances occurred in the study on the long-term toxicity of a Roundup herbicide and a Roundup-tolerant genetically modified maize by Séralini et al. (2012). Following widespread criticism of the article and its purported findings, the journal retracted it in November 2013.

Arguably, a more important challenge to the production of animal food is the question of animal welfare and the increasing demands by consumers that this be

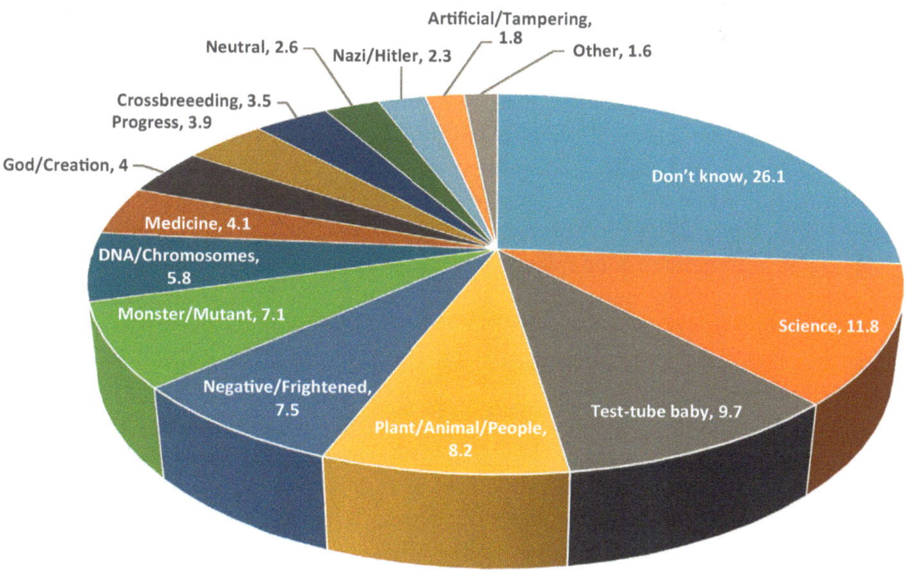

Figure 8 Results (percentage) from respondents who were asked about the first thoughts or images that came to mind when they thought about genetic engineering.
Data from Hallman and Metcalfe (1994).

addressed. To be clear, there is no question that animals should be treated with respect; if for no other reason than pure profit, the evidence shows that well-treated animals perform better than those that are abused (Rushen et al., 1999a,b). Our responsibilities, however, go further than the mere notion of profit. We have a responsibility for the proper use of all of our planet's resources, including those animals that are part of the human food chain. The objective of this chapter is not to argue whether or not we have the right to harvest animals for human purposes. If one holds the view that animals have equal rights to humans and should not be raised for food, there is little to be gained in discussing our responsibility to treat animals in a humane manner when we are routinely ignoring their ultimate right (i.e., their right to life). When asked to comment on this seeming contradiction, Professor Temple Grandin replied that the majority of these animals would never have even existed in the first place if they were not part of the human food chain, and that the least we can do is treat them humanely (personal communication). Assuming we have the right to harvest animals for human needs, the question of welfare often reduces to differences between perception and reality of welfare. The humanization of farm animals is dangerous for the very reason that they are *not* human. At the 2014 Discover Conference on Strategies for Improving U.S. Dairy Cattle Welfare, Ms. LuAnn Troxel (Owner of Troxel Dairy Farm, and Past President and Business Manager, Indiana Dairy Producers, Hanna, Indiana) recounted being once asked "if she loved her cows." Her answer was that she "loved" her children, but that she respected her animals. Anecdotes have little place in a scientific publication except that, in this case, such a response demonstrates the heart of much of the confusion. Animal welfare practices, therefore, should be dictated by the best interest of the animals, rather than some perceived needs by consumers who are unaware of the realities of production systems. Legislation that quibbles over the size of poultry cages versus free-range environments are often oblivious of the potential well-being repercussions for the animals who are not in a controlled environment. McInerney (2004) illustrates this concept in his depiction of welfare ranging from "natural" through "maximal" to "minimal" (Figure 9), the point being that what is perceived as the natural condition for an animal may not represent its optimal welfare state. As an extreme example, a completely free-range hen is susceptible to the fox!

In terms of animal production systems, producers need to be aware of consumer requirements, since failure to account for them leads to discrimination of products. However, an exaggerated humanization of animals has implications for the cost of food production without necessarily achieving the perceived goals of either side. It should be remembered that livestock production not only fulfills the obvious need for the provision of food, but also represents a livelihood for many millions all over the world. While animal welfare is largely recognized as an essential component of the food production chain, unreasonable legislation and restrictions on production systems (that are based on fringe groups or uninformed consumers) may take production out of the hands of those most likely to implement said measures, simply because of profit margins (McGlone, 2001). The result of this would be to push production and increase the prevalence of large-scale systems. Again, much of this debate is taking place in the developed world. One might argue that in parts of the world where death from hunger is a reality every day, such discussions are a luxury. However, this does not negate the

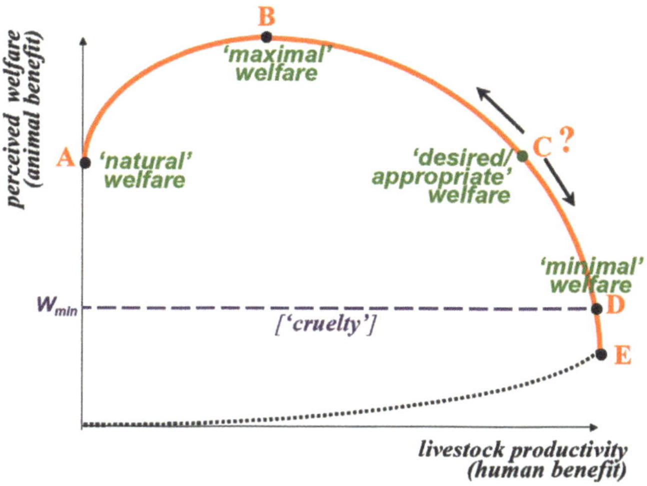

Figure 9 Relationship between livestock productivity (which maps directly into human benefit) on the horizontal axis, and the level of perceived animal welfare on the vertical axis. The point labeled A represents an initial reference point, where no specific management effort is directed toward exploiting the animal's productivity, and it is presumed to experience a particular level of (natural?) welfare. As husbandry inputs are employed to feed and house the animals, protect them from predators, control disease, and so forth, it is generally believed their welfare increases as well as their economic productivity. This is only true up to a point, however (labeled as B), beyond which progressive efforts to exploit biological potential through the application of higher levels of inputs (increasing "intensity of production") achieves rising levels of productivity but at increasing challenge to the animals' perceived welfare. If this process is pursued far enough, it is likely that a point is reached (labeled as E) where the animals can no longer sustain fitness, are literally driven to their biological limits, and the whole production system collapses.
McInerney (2004).

general conclusion that welfare is an animal's right, and could lead some developed countries or regions to discriminate against importing from (developing) countries or regions on the basis of welfare. Fraser (2008) argues that some developing countries could enter into agreements, whereby markets are guaranteed for an implementation of certain welfare standards, and points to the fact that the European Union has already indicated such a willingness.

4.4 Environmental Challenges

The final challenge to the increase in animal products for food comes from their effect on the environment. Again, there is no doubt that animal production has an effect on greenhouse gases; it is, however, a discussion that is fraught with conflicting views, often by groups with specific agendas. Figure 10 from the World Resources Institute shows the relative contributions of various sectors to specific gas emissions. On the one hand, it can be argued that livestock and the resulting manure only contribute a little over 5% of all emissions; on the other hand, one can argue that those specific gases (methane) are more damaging on a per-unit basis than say, CO_2, and that the 5% is, therefore, misleading.

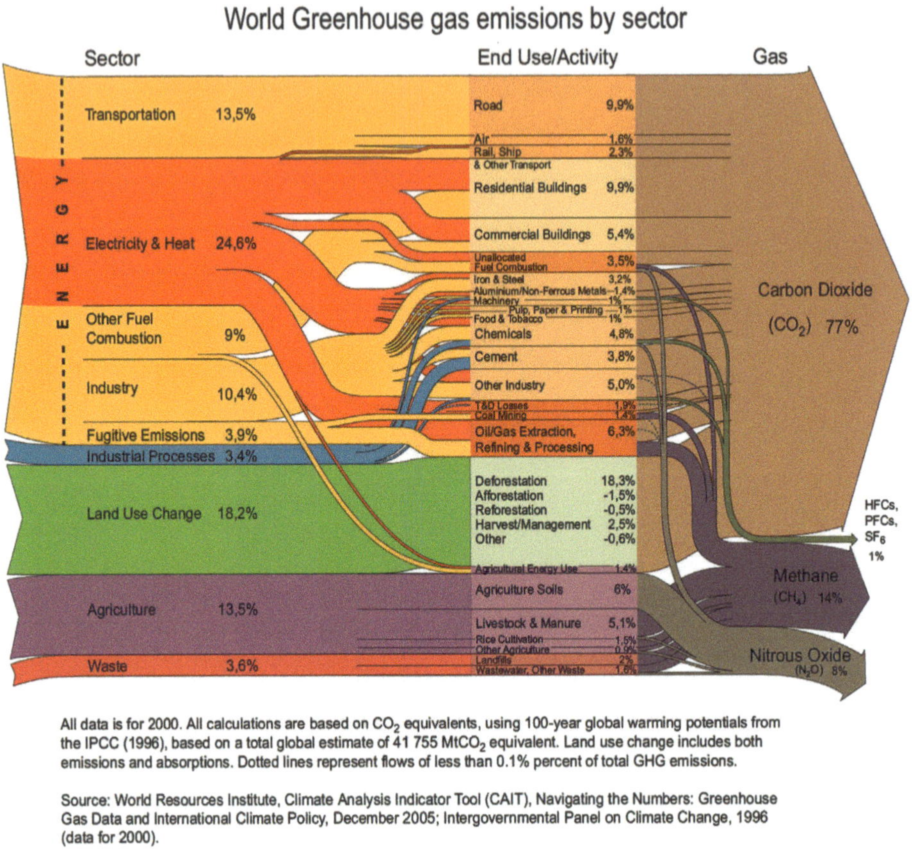

Figure 10 Global greenhouse gas emissions flow chart.
Source: Baumert, K.A., Herzog, T., Pershing, J., 2005. Navigating the Numbers: Greenhouse Gas Data and International Climate Policy. World Resources Institute, http://www.wri.org/sites/default/files/pdf/navigating_numbers.pdf.

One could also debate how much of the costs of transportation and industry should be attributed to livestock production. On the other hand, deforestation (some for arable crop production) and rice production contribute significant amounts in their own right.

The arguments should not be "which produces more?" but rather "how can we minimize the effects?" In fact, animal production has been extremely proactive, as an industry, in limiting its footprint due to all of the technologies discussed in Section 3. The fact that, for example, one cow today can produce the same amount of milk as four cows in the 1940s is an important advance in terms of less water consumption, less manure production, and less land utilization (Capper et al., 2009). Equally impressive savings have been seen in the production of beef (Capper, 2011), where progress over the 10 years in question saw 10% less feed energy, 20% less feedstuffs, 30% less land needed, and 14% less water. In addition, the evidence would suggest that the overall carbon footprint of animal production is decreasing significantly for each unit produced (Capper et al., 2009; Capper, 2011, 2013; Simmons, 2011), although Capper (2013) points to the difficulty of convincing consumers to accept the very practices and technologies that led to this reduction.

The increasing use of life-cycle assessments in agriculture demonstrates a proactive approach by the agricultural industry in acknowledging the need for environmental sustainability and a responsible use of our planet's resources. These approaches are more likely to help with consumer acceptance than reactionary arguments by fringe groups. With more studies looking at life-cycle assessment, it is possible to try and assess the overall environmental impact of specific production systems. Høgaas Eide (2002) actually found that small Norwegian dairy farms had a larger environmental impact than mid- or large-sized ones, while Boggia et al. (2010) found that organic poultry systems had a better environmental performance than conventional systems. The variables are many, and the conclusions often contextual to a specific part of the world, but it shows consumers that the agricultural industries are serious about their footprint and are attempting to minimize it.

Finally, there is an aspect of sustainability to livestock agriculture that goes beyond environmental arguments. It is essential to various sectors by providing a product that is in demand, a viable income to a significant number of the world's smallholders, a manner for using our natural resources efficiently, and a significant contribution to human health (FAO, 2014).

5. Future Trends

The author Rick Cook (1989) has been quoted as saying:

> *Programming today is a race between software engineers striving to build bigger and better idiot-proof programs, and the universe trying to produce bigger and better idiots. So far, the universe is winning.*

While this statement was made, somewhat tongue-in-cheek, regarding the development of software, this may be equally applicable with regard to our efforts to feed a growing global population. Scientists strive to improve the ability to meet societal needs through the use of technology that has, for the most part, been vetted via rigorous experimentation. Despite these many advances, the developed world seems to focus sometimes on the trivial and sometimes on the ideas of fringe groups that are coming from a very privileged place. As access to, and ability to purchase food (animal or plant) becomes less of a challenge, we seem to place importance on details that are incredulous to other parts of the world where almost 1 billion are undernourished (FAO, 2011b). In North America, we live in a society where, too often, pseudoexperts with no specific scientific training have found gullible audiences that will listen to them, simply because they are famous. Some of the scientific community finally spoke out when a Hollywood star promoted a cookbook that was accused of bordering on quack science (Maclean's, April 2013). It is frustrating to know that, even in the developed world, 2 billion suffer from at least one micronutrient deficiency, 1.4 billion are estimated as overweight (with 500 million obese), and 26% of world's children are stunted (FAO, 2011b). It is disingenuous to have people in a position of influence preaching on areas about which they have no educational qualifications.

Simmons (2011) calls for a truce regarding the safety of technology in the role of food production and advocates three rights: every human's right to have access to affordable food, the consumer's right to spend his or her food budget on the widest variety of food choices, and the creation of sustainable food production systems that are environmentally right. In other words, mankind should have access to a wide variety of affordable food that has been produced with regard to our environment. The FAO (2011b) would seem to agree: they advocate the four pillars of food security as availability, access, stability, and utilization.

Simmons (2011) also quoted results from The International Consumer Attitudes Study (ICAS) to conclude that 99% of global consumers make their purchase decisions based on taste, cost, nutrition, and some level of lifestyle choice. This is important to keep in mind when we read headlines about the "next killer ingredient in our diets," or we see news reports about fringe groups protesting the eating of animal products; they are by far in the minority.

It is clear that we will need more food than ever (animal and plant products) as we approach the middle of this century. It is also clear that technology can play an enormous role in bringing this about within a framework of sustainability and respect for our planet's resources. It is now the task of scientists, industry, and consumers to do a better job of promoting these rigorously tested methodologies to a public that not only wants to know what it is eating but has a right to that information in a non-confrontational or defensive manner. As the majority of the world's population continues to become even more urban, it is imperative that those sectors—furthest removed from agriculture and the production of food—be made aware of where their food comes from, how it is produced and processed, and what are the nutrient contents. It is also clear that developed countries will need to give serious thought to the amount of food that is wasted on an annual basis, and how such waste can be minimized, especially in societies that pay double the amount for a bottle of water over a bottle of soda.

References

Alexandratos, N., Bruinsma, J., 2012. World Agriculture towards 2030/2050: The 2012 Revision. ESA Working paper No. 12-03. FAO, Rome.

Benchaar, C., Petit, H.V., Berthiaume, R., Whyte, T.D., Chouinard, P.Y., 2006. Effects of dietary addition of essential oils and monensin premix on digestion, ruminal fermentation characteristics, milk production, and milk composition in dairy cows. Journal of Dairy Science 89, 4352–4364.

Bender, A., 1992. Meat and Meat Products in Human Nutrition in the Developing World. Food and Nutrition paper No. 53 FAO, Rome. http://www.fao.org/docrep/t0562e/t0562e00.HTM.

Boggia, A., Paolotti, L., Castellini, C., 2010. Environmental impact evaluation of conventional, organic and organic-plus poultry production systems using life cycle assessment. World's Poultry Science Journal 66, 95–114.

Briggs, H.M., Briggs, D.M., 1980. Modern Breeds of Livestock, fourth ed. Macmillan Publishing Co.

Canada's Food Guide, 2015. Health Canada. http://www.hc-sc.gc.ca/fn-an/food-guide-aliment/basics-base/1_1_1-eng.php (accessed March 2015).

Canadian Dairy Information Centre, 2014. http://aimis-simia-cdic-ccil.agr.gc.ca/rp/index-eng.cfm?action=rR&pdctc=&r=213 (accessed 04.03.15.).

Canadian Dairy Network, Beavers, L., Van Doormaal, B., 2013. Lifetime Profit Index: Are We Achieving Our Goals? http://www.cdn.ca/document.php?id=336 (accessed March 2015).

Canadian National Farm Animal Care Council. http://www.nfacc.ca/codes-of-practice.

Capper, J.L., 2011. The environmental impact of beef production in the United States: 1977 compared with 2007. Journal of Animal Science 89, 4249–4261.

Capper, J.L., 2013. Should we reject animal source foods to save the planet? A review of the sustainability of global livestock production. South African Journal of Animal Science0375-1589 43 (3), 233–246 [online].

Capper, J.L., Cady, R., Bauman, D., 2009. The environmental impact of dairy production: 1944 compared with 2007. Journal of Animal Science 87, 2160–2167.

Carter, C.A., Gruère, G.P., 2003. Mandatory labeling of genetically modified foods: does it really provide consumer choice? AgBioForum. 6 (1&2), 68–70. www.agbioforum.org/v6n12/v6n12a13-carter.htm.

Carter, C.A., Gruère, G.P., McLaughlin, P., MacLachlan, M., 2012. California's proposition 37: effects of mandatory labeling of GM food. Agricultural and Resource Economics Update. 15 (6), 3–8. http://giannini.ucop.edu/media/are-update/files/articles/V15N6_2.pdf.

Conte Jr., R.L., 2015. The Global Food Crisis: Feeding Ten Billion Persons: An Analysis of World Staple Crops by Macronutrients. (accessed March 2015) http://www.gardening-places.com/articles/global-food-crisis.htm.

Cook, R., 1989. The Wizardry Compiled. Baen Books, New York, ISBN: 0671698567.

Curtis, K.R., McCluskey, J.J., Wahl, T.I., 2004. Consumer acceptance of genetically modified food products in the developing world. AgBioForum 7 (1&2), 70–75.

Dechow, C., May 25, 2014. Bias remains a challenge for genomic evaluations. Hoard's Dairyman 373.

DePeters, E.J., Cant, J.P., 1992. Nutritional factors influencing the nitrogen composition of bovine milk: a review. Journal of Dairy Science 75, 2043–2070.

DiLorenzo, N., 2011. Manipulation of the rumen microbial environment to improve performance of beef cattle. In: 22nd Florida Ruminant Nutrition Symposium. February 1–2, 2011http://dairy.ifas.ufl.edu/rns/2011/11diLorenzo.pdf (accessed March 2015).

Dürr, J., Philipsson, J., January 2012. International cooperation: the pathway for cattle genomics. Animal Frontiers 2 (1). http://dx.doi.org/10.2527/af.2011-0026.

Eastridge, M.L., 2006. Major advances in applied dairy nutrition. Journal of Dairy Science 89, 1311–1323.

Eisler, M.C., Lee, M.R.F., Tarlton, J.F., Martin, G.B., Beddington, J., Dungait, J.A.J., Greathead, H., Liu, J., Mathew, S., Miller, H., Misselbrook, T., Murray, P., Vinod, V.K., Van Saun, R., Winter, M., March 6, 2014. Agriculture: steps to sustainable livestock. Nature 507 (7490), 32–34.

FAO, WFP, IFAD, 2012. The State of Food Insecurity in the World 2012. Economic Growth is Necessary but Not Sufficient to Accelerate Reduction of Hunger and Malnutrition. FAO, Rome.

FAO, Abberton, M.T., MacDuff, J.H., Marshall, A.H., Humphreys, M.W., 2008. The Genetic Improvement of Forage Grasses and Legumes to Enhance Adaptation of Grasslands to Climate Change. Institute of Grassland and Environmental Research, Aberystwyth, UK, in collaboration with Plant Production and Protection Division Crop and Grassland Service of the Food and Agriculture Organization of the United Nations. http://www.fao.org/3/a-ai779e.pdf.

FAO, 2011a. Global Food Losses and Food Waste – Extent, Causes and Prevention. Rome.
FAO, 2011b. World Livestock 2011 – Livestock in Food Security. FAO, Rome.
FAO, 2012a. Balanced Feeding for Improving Livestock Productivity – Increase in Milk Production and Nutrient Use Efficiency and Decrease in Methane Emission, by Garg, M.R. FAO Animal Production and Health Paper No. 173. Rome, Italy.
FAO, 2012b. Impact of Animal Nutrition on Animal Welfare – Expert Consultation 26–30 September 2011. FAO Headquarters, Rome, Italy. Animal Production and Health Report. No. 1. Rome.
FAO, 2012c. Livestock Sector Development for Poverty Reduction: An Economic and Policy Perspective – Livestock's Many Virtues, by J. Otte, A. Costales, J. Dijkman, U. Pica-Ciamarra, T. Robinson, V. Ahuja, C. Ly and D. Roland-Holst. Rome, p. 161.
FAO, September 29–October 3, 2014. Committee on Agriculture – 24th Session. The Global Agenda for Sustainable Livestock, Rome. http://www.fao.org/3/a-ml052e.pdf.
Field, T.G., Taylor, R.E., 2010. Scientific Farm Animal Production, Paperback – December 30, 2010, tenth ed. Prentice Hall.
Foote, R.H., 2002. The history of artificial insemination: selected notes and notables. Journal of Animal Science 80 (2), 1–10. http://dx.doi.org/10.2134/animalsci2002.80E-Suppl_21a.
Fraser, D., 2008. Toward a global perspective on farm animal welfare. Applied Animal Behaviour Science 113, 330–339.
Garner, D.L., Seidel Jr., G.E., 2008. History of commercializing sexed semen for cattle. Theriogenology 69 (7), 886–895. http://dx.doi.org/10.1016/j.theriogenology.2008.01.006.
Goddard, M.E., Hayes, B.J., 2009. Mapping genes for complex traits in domestic animals and their use in breeding programmes. Nature Reviews Genetics 10 (6), 381–391.
Gruère, G.P., Rao, S.R., 2007. A review of international labeling policies of genetically modified food to evaluate India's proposed rule. AgBioForum. 10 (1), 51–64. http://agbioforum.org/v10n1/v10n1a06-gruere.htm.
Hallman, W.K., Metcalfe, J., 1994. Public Perceptions of Agricultural Biotechnology: A Survey of New Jersey Residents. Food Policy Institute, Cook College, Rutgers. The State University of New Jersey. Available at: www.foodpolicyinstitute.org.
Halweil, B., May/June 2006. Can organic farming feed us all? World Watch Magazine. 19 (3). http://www.worldwatch.org/node/4060.
Hayes, B.J., Bowman, P.J., Chamberlain, A.J., Goddard, M.E., 2009. Invited review: genomic selection in dairy cattle: progress and challenges. Journal of Dairy Science 92 (2), 433–443. http://dx.doi.org/10.3168/jds.2008-1646.
Hoban, T.J., 1997. Consumer acceptance of biotechnology: an international perspective. Nature Biotechnology 15, 232–234.
Høgaas Eide, M., 2002. Life cycle assessment of industrial milk production. The International Journal of Life Cycle Assessment 7, 115–126.
Jenkins, T.C., McGuire, M.A., 2006. Major advances in nutrition: impact on milk composition. Journal of Dairy Science 89 (4), 1302–1310.
Lock, A.L., Bauman, D.E., 2004. Modifying milk fat composition of dairy cows to enhance fatty acids beneficial to human health. Lipids 39, 1197–1206.
Maclean's, April 2, 2013. Julia Belluz on 'It's All Good' and the Dangers of Elimination Diets.
McGlone, J.J., 2001. Farm animal welfare in the context of other society issues: toward sustainable systems. Livestock Production Science 72, 75–81.
McInerney, J., 2004. Animal Welfare, Economics and Policy: A Report on a Study Undertaken for the Farm & Animal Health Economics Division of the Department for Environment. Food & Rural Affairs, UK. http://archive.defra.gov.uk/evidence/economics/foodfarm/reports/documents/animalwelfare.pdf.

Millar, M., June 22, 2010. Where's the Pork? Financial Sense Observations Archive. http://www.financialsensearchive.com/Market/millar/2010/0622.html (accessed March 2015).
Mitchell, P., 2003. Europe sees sharp decline in GMO research. Nature Biotechnology. 21, 468–469. http://dx.doi.org/10.1038/nbt0503-468. http://www.nature.com/nbt/journal/v21/n5/full/nbt0503-468.html.
Moon, W., Balasubramanian, S.K., 2001. Public perceptions and willingness-to-pay a premium for non- GM foods in the us and UK. AgBioForum 3 (4), 221–231.
National Agricultural Biotechnology Council (NABC) Report 11 1999. World Food Security and Sustainability: The Impacts of Biotechnology and Industrial Consolidation. Boyce Thompson Institute, Ithaca, NY 14853.
National Center for Biotechnology Information. http://www.ncbi.nlm.nih.gov/genome.
National Research Council (United States). http://www.nationalacademies.org/publications/index.html.
Nutrient requirements of dairy cattle/Subcommittee on Dairy Cattle Nutrition, Committee on Animal Nutrition, Board on Agriculture, National Research Council. — 7th rev. ed.
Parfitt, J., Barthel, M., Macnaughton, S., 2008. Food waste within food supply chains: quantification and potential for change to 2050. Philosophical Transactions of the Royal Society B (2010) 365, 3065–3081.
de Roos, A.P.W., Schrooten, C., Veerkamp, R.F., van Arendonk, J.A.M., 2011. Effects of genomic selection on genetic improvement, inbreeding, and merit of young versus proven bulls. Journal of Dairy Science 94 (3), 1559–1567. http://dx.doi.org/10.3168/jds.2010-3354.
Rushen, J., Taylor, A.A., de Passillé, A.M., 1999b. Domestic animals' fear of humans and its effect on their welfare. Applied Animal Behaviour Science 65 (3), 285–303.
Rushen, J., De Passillé, A.M., Munksgaard, L., 1999a. Fear of people by cows and effects on milk yield, behavior, and heart rate at milking. Journal of Dairy Science 82 (4), 720–727.
Schaeffer, L.R., 2006. Strategy for applying genome-wide selection in dairy cattle. Journal of Animal Breeding and Genetics 123, 218–223.
Seidel Jr., G.E., August 2007. Overview of sexing sperm. Theriogenology 68 (3), 443–446. http://dx.doi.org/10.1016/j.theriogenology.2007.04.005.
Séralini, G.-E., Clair, E., Mesnage, R., Gress, S., Defarge, N., Malatesta, M., Hennequin, D., Spiroux de Vendômois, J., 2012. Long term toxicity of a Roundup herbicide and a Roundup-tolerant genetically modified maize. Food and Chemical Toxicology 50, 4221–4231.
Séralini, G.-E., Clair, E., Mesnage, R., Gress, S., Defarge, N., Malatesta, M., Hennequin, D., Spiroux de Vendômois, J., 2012. Long term toxicity of a Roundup herbicide and a Roundup-tolerant genetically modified maize. Retraction http://www.sciencedirect.com/science/article/pii/S0278691512005637.
Seufert, V., Ramankutty, N., Foley, J.A., May 10 , 2012. Comparing the yields of organic and conventional agriculture. Nature 485, 229–232. http://dx.doi.org/10.1038/nature11069.
Simmons, J., 2010. Technology's role in the 21st century: food economics and consumer choice: why agriculture needs technology to help meet a growing demand for safe, nutritious and affordable food. (AGR: IND44331254). Proceedings – Cornell Nutrition Conference for Feed Manufacturers 2010, 171–185.
Simmons, J., 2011. Making Safe, Affordable and Abundant Food a Global Reality. Elanco Animal Health White Paper AI11192.
Smith-Spangler, C., Brandeau, M.L., Hunter, G.E., Bavinger, J.C., Pearson, M., Eschbach, P.J., Sundaram, V., Liu, H., Schirmer, P., Stave, C., Olkin, I., Bravata, D.M., 2012. Are organic foods safer or healthier than conventional alternatives? a systematic review. Annals of Internal Medicine 157, 348–366.

Teisl, M.F., Garner, L., Roe, B., Vayda, M.E., 2003. Labeling genetically modified foods: how do us consumers want to see it done? AgBioForum 6 (1&2), 48–54.
Troxel, L., 2014. The importance of dairy welfare for our industry. In: 27th American Dairy Science Association's Discover Conference on Food Animal Agriculture: Strategies for Improving US Dairy Cattle Welfare (May 27–30, 2014, Itasca, IL). http://www.adsa.org/Portals/0/SiteContent/docs/discover/DC27/DC27_Speaker_Interpretive_Summaries.pdf.
United Nations, 2014. Department of Economic and Social Affairs, Population Division. World Urbanization Prospects: The 2014 Revision, Highlights (ST/ESA/SER.A/352).
USDA Economic Research Service Data. http://www.ers.usda.gov/data-products/international-agricultural-productivity.aspx (accessed October 2015).
de Vries, A., Overton, M., Fetrow, J., Leslie, K., Eicker, S., Rogers, G., 2008. Exploring the impact of sexed semen on the structure of the dairy industry. Journal of Dairy Science 91, 847–856.
Wikipedia, 2015. http://en.wikipedia.org/wiki/Genetically_modified_organism (accessed 06.08.15.).

Aquaculture Technologies for Food Security

5

David C. Little[1], *Stuart W. Bunting*[2]
[1]Institute of Aquaculture, University of Stirling, Stirling, UK; [2]Bunting Aquaculture, Agriculture and Aquatic Resources Conservation Services, Glemsford, Suffolk, UK

1. Introduction

Aquaculture has been identified as a critical part of supporting food security, especially for low- and medium-income countries (LMIC) in which fish is an established and key part of diets (Béne et al., in press). Finfish and other aquatic products ("fish") are high in protein and rich in micronutrients. Employment for low-income people throughout aquaculture value chains is increasingly contributing both directly and indirectly to their food and nutritional security (Bostock et al., 2010).

Technology[1] related to aquaculture has been defined both narrowly and more broadly, but its role in delivering enhanced food security demands a holistic approach that encompasses the complete value chain.[2] For example, the principle "technology" for production might be simple, such as earthen ponds as culture units, whereas that required for producing appropriate feeds or processing shrimp or fish to a value-added product would typically be much more complex.

Food security rebalances the equation between production and consumption through a long-negotiated (Maxwell, 1996), but now generally accepted, definition of "when all people at all times have access to sufficient, safe, nutritious food to maintain a healthy and active life." This raises the issue of the degree to which aquaculture in its different forms contributes to the food security of different groups and through which channels. The level to which smallholder households might produce aquatic foods for self-sufficiency, compared to using cash from sales to support consumption indirectly, is one question. The types of technologies most appropriate for different strategies will largely depend on context and the broader strategies that households use to maintain their livelihoods (Bunting et al., 2015). The growth in aquaculture, however,

[1] The broad definition of technology adopted here includes equipment, infrastructure, and machines (e.g., computers, engines, pumps, vehicles), biotechnology (e.g., diagnostic tests, ecological engineering, feed formulation, biomarkers for genetic selection and broodstock management, probiotics, vaccines), and knowledge and skills (e.g., better management practices, biosecurity, human capital, protocols, and techniques) to enhance productivity, optimize resource use efficiency, minimize production risks, avoid post-harvest losses, and mitigate environmental impacts.

[2] The assessment presented here considers the emergence of technology across product value chains from production-enhancing inputs (broodstock and seed, feed, fertilizer), production and husbandry, harvest and processing, distribution, sales and marketing, and waste disposal and recycling.

Emerging Technologies for Promoting Food Security. http://dx.doi.org/10.1016/B978-1-78242-335-5.00005-6

has largely been a market-led phenomenon, and commercial enterprises ranging from family-run businesses to corporate investments have been the main drivers in what are now recognized as innovation systems (Klerkx et al., 2012). Development and investments away from the farm, through the value chain as a whole, have been critical to ensure delivery of aquaculture products at a price and quality that consumers can afford. Any assessment of impacts on food security arising from aquaculture needs to recognize this complexity.

Affordability and access to seafood for poorer groups might be in doubt, given its comparatively high value compared to alternatives, and its integration into global trade. Technology development will be only one of several changes required to ensure global food security (Garnett et al., 2013). The impacts of changing consumer perceptions and demand for aquatic products will be critical, but so will the terms of global trade between powerful and weaker players. The interactions between aquaculture emerging as a major source of food and the impact of this growth on aquatic environments and wild fish stocks they support has stirred considerable debate. The processing of lower-value species into "marine ingredients" in aquafeeds, for example, has been criticized for squeezing availability of fish as direct food for poorer people (Naylor et al., 2000; Muir, 2005), although such views have been criticized as simplistic by others (e.g., FAO, 2002).

The pioneer phase of aquaculture development has been characterized by its use of a wide range of species cultured. In the short to medium term, this is likely to decline, as production consolidates around fewer species in which investments in nutrition, breeding, and health management can be focused to optimize productivity and profitability. This could lead to food security benefits being lost unless steps are taken to screen and evaluate available species for their nutritional and cultural value. One example of this has been the efforts to commercialize the integration of a native local species, mola (*Amblypharyngodon mola*), into carp polycultures in Bangladesh (Wahab et al., 2011). This species was prioritized based on research showing both its particularly high levels of available vitamin A and other micronutrients lacking in human diets and the high cultural acceptability and value of the species (Roos et al., 2007; Thilsted, 2012).

Development of aquaculture is having impacts on broader food security through linking resource use, both on the farm and between different nodes of the value chain. Diversification of terrestrial livestock feed production and health management to the rapidly growing aquaculture sector is one obvious example, but ponds as water reservoirs can have extensive multiplier effects on food production at the farm and community level.

In the following sections, we first consider aquaculture in terms of broader foodscapes before considering the key drivers of change. We then consider the options for sustainable intensification and its limitations based on the energy, water, and nutrient nexus that different designs impose, before considering how the life stage of cultured animals affects technological choices and reflecting on issues of health management and genetic gain. We conclude by considering trajectories of the sector toward both specialization and integration and the implications for food security.

2. Aquaculture, Fisheries, and Related Forms of Food Production

Major imperatives in the global foodscape for promoting aquaculture as a mechanism to improve food security has been the widespread decline in supplies from wild fishery stocks and the nutritional importance of fish and other aquatic products (FAO, 2014). Poorer countries often rely more on fish and other aquatic food than better-off countries for their nutritional security. Interventions made by international and national organizations have often sought to promote aquaculture as a component of smallholder livelihoods based around a "simple" technology approach aiming to meet household subsistence needs. Such initiatives, particularly if limited to pond excavation, have often failed or had little sustained impact after the withdrawal of subsidies. Various explanations have been advanced for this failure, but the lack of appreciation of constraints to poorer smallholders and the roles aquaculture might play in complex livelihoods have often been central.

As with technological packages developed to propagate the Green Revolution in agricultural production in the 1940s to 1960s, there is a danger that aquaculture technologies will be promoted where they are inappropriate, given the prevailing environmental conditions or sociocultural setting. The formulation of suitable recommendation domains could help better target uptake promotion initiatives, and producers adopting new technologies (in the broadest sense) should be supported with appropriate capacity-building and enabling institutional arrangements. Technological developments should, ideally, be demand-led according to the needs and capacities of prospective user groups, and research and development support should be targeted at the joint assessment of constraints to food security and opportunities for innovation with producers and other stakeholders in the value chain. Costs, both capital and operating, and environmental and socioeconomic impacts of adopting aquaculture technologies, must be included in assessments and risks associated with uptake, as well as demand considerations and the implementation of appropriate safeguards. Any evaluation of aquaculture technologies to safeguard food security must take into account the scale at which particular interventions might be able to respond to trends associated with typical aquaculture development (Table 1).

Even if food fish production itself is "simple" in terms of technology, it may be dependent on linkages with more technically demanding seed and feed provision. Where aquaculture has become established, it has often been built on resources and capacity already established within fishery value chains; this may relate to equipment such as nets and boats, or market infrastructure such as ice plants and cold storage. Commercial actors throughout often complex value chains have been essential to aquaculture spreading in impact and significance in markets, both local and distant to production centers. Aquaculture almost certainly contributes to food security of poorer people, mainly through employment and consumption rather than ownership or even operation of a food production enterprise (Belton et al., 2014). Opportunities occur throughout aquaculture value chains both upstream and downstream, and often with low barriers to entry. Processing of farmed fish typically remains a labor-intensive activity, despite sophisticated fish-processing technology being available. In countries where aquaculture delivers a significant proportion of the fish consumed, in general, prices in the market have moderated and fish have become more affordable (Little et al., 2012) to poorer consumers.

Table 1 Opportunities for Aquaculture Technologies to Promote Food Security at Multiple Scales across the Driving forces, Pressures, State, Impacts, Responses (DPSIR) Cause-Effect Continuum

Cause–Effect Continuum Stage	Description	Responses
Driving forces	• Globalization and competition internationally. • Economic development and industrialization. • Population growth, urbanization, and more affluent consumers.	• Governments must be responsive to food shortages and commodity price spikes to ensure poor and marginal populations can meet their basic needs for survival. • International and regional organizations should monitor production and trade statistics to preempt food security problems. • Regional and national development plans should promote the adoption of appropriate technologies to ensure aquaculture development is efficient and equitable.
Pressures	• Demand for higher-value aquaculture products increased owing to increased trade internationally and a burgeoning population of middle-class consumers nationally. • Demand for land for aquaculture development leads to conversion of agricultural land and natural areas, particularly wetlands. • Commodity and labor costs increased owing to higher demand for inputs globally and economic development prompting wage growth. • Producers switch from the semi-intensive culture of mixed species to monocultures of more valuable species destined for export and burgeoning urban markets.	• Promote culture systems combining intensive production of cash-crop species and lower value integrated or foraging species to meet dual economic development and food security objectives and reduce demands placed on supporting ecosystem areas, principally in regard to waste assimilation. • Protect ecosystems, notably wetlands that sustain stocks and flows of ecosystem services that support aquaculture production and food production in interconnected agroecosystems. • Promote integrated aquaculture–agriculture production systems to optimize resource use efficiency and financial returns, and safeguard staple food supplies.

State	• Aquaculture produces cash-crops that are transported to centralized processing facilities that meet national or international standards before being sold by urban retailers or exported. • Intensive production of a single species that is dependent on formulated feed inputs regarded as having the greatest potential to maximize financial returns for producers. • Businesses become consolidated and vertically integrated to capitalize on economies of scale and reduce transaction costs.	• Adopt better management practices to optimize production, harvest scheduling, and optimize processing and distribution arrangements to reduce losses and increase the shelf life of products. • Support SME producers with training and capacity building to reduce risks, promote production efficiency, and create self-sustaining businesses that provide employment opportunities. • Foster clusters and groups of farmers and promote the formation of producer organizations to enhance knowledge exchange and policy-engagement, and strengthen research and development capacity and funding.
Impacts	• Production of lower-value species ceases, and the availability of affordable fish in local markets declines. • Conversion of agricultural land and natural areas to aquaculture reduces the availability of staple crops locally, and loss of access to common property resources that sustain provisioning ecosystem services. • Production reliant on formulated feeds manufactured elsewhere, and the centralized processing of products results in the loss of employment opportunities locally. • Intensification of production and the expansion of culture systems can exceed the carrying capacity of supporting ecosystem areas to assimilate wastes and mitigate other environmental impacts.	• Subsidize small-scale aquaculture producers with production-enhancing inputs (quality seed and feed and fertilizer) to bolster local food supplies in preparation for food-insecure periods, especially in the face of anticipated worsening climate-change-induced impacts. • Protect natural areas that sustain stocks and flows of ecosystems services that support aquaculture production, but permit continued access for wise use for food security among poor and marginal communities. • Limit the extent of aquaculture development so that it does not exceed the environmental carrying capacity, and where possible use ecological engineering approaches to reduce dependence on the appropriation of ecosystem services.

3. Key Change Drivers

The key drivers of technical change remain increasing demand for farmed aquatic products coupled with the need for greater productivity. Ever more formalized governance demands—that increased production comes with higher levels of assured public safety and reduced environmental impacts—have evolved in recent years, particularly in high-income countries. A range of other so-called credence qualities also affect value and access to certain markets (Fernquist and Ekelund, 2014). In general, better-off consumers lead the interest in credence qualities (e.g., organic, fair trade, "green" status), but are also likely to demand fish of better eating quality based on taste, size, and freshness. Increasingly private standards are supplementing government regulation, a process that is partly driven by the competitive and brand-protective retailers and food service operators in developed country markets (Bush et al., 2013).

Globally, increased population growth and rising purchasing power is fueling demand, particularly in areas of the world where seafood constitutes a key part of diets. When expressed as the proportion of total protein consumption, large areas of West and Central Africa and South and Southeast Asia emerge as key areas of likely growth. Traditionally, supply and demand for whole fresh fish has often been part of a broader array of dried, salted, and fermented products that have been traded at distance. Migration of fish-eating people has often been a driver for the development of trade opportunities between centers of production and consumption.

Increasing urbanization tends to raise demand for value-added and processed products. This has often led to better quality control in market chains (more icing, refrigeration) and to live marketing. Such changes have implications for postharvest capacity, including technology, at both enterprise and broader levels. In addition to the marketing format, such social changes are also likely to change the types of aquatic animal in greatest demand. Upgrading herbivorous species through value addition may prove essential to maintain them in the diets of increasingly urban, time-short populations, but will require a range of different downstream technological innovations. Removal of fine intramuscular bones and development of products, such as surimi, based on freshwater fish, have already been piloted and show great promise. The spread and universal popularity of ready-to-eat format food in general, as supermarket culture spreads, is noteworthy (e.g., Zhou et al., 2015), as is the rise in consumption outside the home (i.e., food service) in terms of overall consumption.

Reliable electrical and potable water supplies and the quality of roads will affect the likely options available, in turn affecting the distance over which fish can be profitably marketed. International trade has been facilitated by growth in air, and especially sea, freight of processed products. Specialized technology, evolving from the much larger trade in processed wild catch, depends on good preharvest, harvest, transportation, and slaughter procedures, but can result in premium products because of the greater consistency possible for cultured compared to wild stocks.

Food and nutritional security through aquaculture may also be impacted by trends toward greater intensification of farming, including the replacement of marine ingredients with lower-cost ingredients of terrestrial origins. The nutritional value of farmed seafood, particularly with regard to levels of N-3 fatty acids, reflects the composition of the diets used, the broader production systems, and the species produced (Karapanagiotidis et al., 2006). Increasing costs of production, especially those relating to land and labor, are major incentives for intensification of production, and the fundamental technology options will now be considered.

4. Intensification: Options and Implications

Sustainable intensification of food production in general requires technical innovation to attain efficiency gains. Waste reduction can be a key component of this process and one critical to working within the so-called energy, water, and nutrient nexus (Beck and Villaorel Walker, 2013). This is particularly challenging for aquaculture compared to terrestrial livestock. Achieving productivity gains in aquatic systems through use of higher stocking densities and feeding intensities, such as has been achieved in terrestrial feedlot livestock systems, can quickly exceed the carrying capacity of the system.[3] Declines in water quality within the culture environment lead to retardation of individual growth of the aquatic animals being farmed, their survival, and as a consequence, overall productivity. Balancing feed availability and use by the cultured animal and the removal of waste is a key element in maintaining productive systems. Aquatic feedlot systems, where all feed requirements are supplied by exogenous feed, are of increasing importance globally. Although knowledge of nutritional requirements lags that of the key terrestrial livestock by some margin, there has been major progress in developing high-quality diets for the key farmed aquatic species. Feed formulation has been particularly critical for carnivorous species, such as the Atlantic salmon, which can only economically be farmed intensively, since natural food webs can only support limited productivity. In contrast, lower trophic species that can grow on aquatic pastures of natural feed (plankton, benthos, etc.) have often been produced more extensively, through stimulation of natural food productivity and only supplementation with readily available and low-cost feeds.

In general, there have been incentives to intensify production, regardless of natural trophic status in the earthen ponds and cages suspended in large water bodies and coastal waters that produce most of the global crop. The key objectives to maintaining water quality are ensuring dissolved oxygen remains above, and nitrogenous compounds (unionized ammonia and nitrite) below, critical levels. Simplistically, while cages rely on dilution with water in the wider water body to maintain adequate water quality, pond managers can employ a range of tactics including aeration and water exchange. Removal of solids is a key part of maintaining water quality, and densely stocked cages and ponds can soon encounter problems if these accumulate within or close to the fish production unit. In all but the most intensively managed ponds, such solids that build up as sediments can be removed periodically, usually after harvest of the pond and water drainage. For cage culture, densities of cages and fish need to be maintained within the local carrying capacity[3] of the larger water body.

[3] The ability of the water to support net growth of the cultured stock.

Inevitably, pressures to intensify in both systems often tend to push the envelope, resulting in deterioration in water quality and resultant system productivity (and profitability). For ponds this can be the outcome of an individual pond operator trying to stock fish at too high a density and feeding at levels that cannot be sustained, given the quality and quantity of water available. This may be exacerbated where pond aquaculture occurs in geographical clusters and many operators are dependent on shared water supplies. Overconcentration of many cages in common property lakes and reservoirs leads to the same outcomes. Such "tragedy of the commons" tends to undermine the role of aquaculture in improving food security and can also undermine associated livelihoods in fishing and service industries.

Solutions to such trajectories of intensification encompass both technological and management components, and both have direct livelihood impacts on those involved. For example, governance that imposes production limits through zonation of production may be one approach (Brummett, 2013). This may result in exclusion of less powerful producers and have disproportionately negative impacts on poorer people. Larger, better-capitalized, cage-based producers may be able to relocate to larger more exposed sites where water quality can be maintained, for example.

Improved feeds and feeding management that lead to less waste is another approach,[4] and many certification schemes demand product that has been raised at above-threshold feed efficiencies. This type of requirement may restrict or remove opportunities for local sourcing and processing of feeds, with negative impacts on local actors employed in such activities.

Further intensification requires cost-effective solids removal as a key design feature. Tanks and raceways utilize water and air flows to concentrate solids in central or peripheral sumps to facilitate their removal and disposal, but this is more problematic in earthen ponds. Nevertheless, improved designs based on central drainage that allow accumulation and regular flushing out of solids, and strategic water movement using airlifts or other forms of aeration, can be effective, particularly in smaller ponds (<1000 m^2). Pond liners can further increase efficiencies of such systems. Farmers innovating super-intensive river catfish ponds in Vietnam utilize simple approaches for raising this tolerant, air-breathing fish at densities in excess of 500 MT/ha (Phuong and Oanh, 2010). The very deep ponds (up to 6 m) act as sedimentation basins, and sludge is manually pumped out two or three times in a single 6-month production cycle (Pham, 2014). Such designs therefore provide local employment opportunities and support food security in a very direct way.

Strategies for the profitable on-site use of such nutrient-rich effluents in surrounding horticulture might also be a powerful incentive for their reuse, but issues of scale are significant hurdles in this respect (Pham, 2014). Such pond-based strategies have tended to evolve in water-rich environments where water exchange is not costly or subject to onerous external regulation, but integrating aquaculture into more

[4] Low-volume, high-density cage culture technology is being actively promoted for growing fish in both freshwater and brackish/marine environments in Asia (Edwards, 2012). The use of high-quality pelleted feed is advocated to optimize the conversion of nutrients to fish and hence minimize adverse environmental impacts.

water-challenged economies has necessitated further intensification. Regulations promoting lower farm effluents have also been a powerful spur to intensify aquaculture system to the point of zero or highly limited emissions (Bergheim and Brinker, 2003).

Limited water availability and/or suboptimal ambient temperature regimes have also incentivized the development of recirculated aquaculture systems (RAS).

The energy/nutrient/water nexus reflects the challenge of increasing productivity in terms of water used through optimal nutrition of fish held at high densities. Improving the water stability, digestibility, and "fit" with nutritional requirements of the fish is a key part of the technological challenge toward improved feeds, as is the energy required to maintain optimal water quality. System filtration design and optimal pumping efficiencies of intensive RAS are critical in this respect, but the availability of consistent, low-cost electrical power is the main factor in determining their likely commercial take-up. Autonomous, solar, or tidal approaches appear to have the most potential going forward, suggesting their likely spatial concentrations in sunny and coastal locations, respectively. Currently, these approaches are mainly taken up by value-added parts of aquaculture—ornamentals and juvenile production or for very high-value species that perform poorly in open-cage or pond systems, e.g., marine flatfish such as turbot. Reliance on natural feed, in general, diminishes with intensification, since management of chaotic and unpredictable natural food webs while sustaining consistent and optimal nutrition and water quality is problematic. This includes biofloc systems (Little et al., 2008), but such production strategies may have other advantages, such as health protection through probiotic effects. The costs of managing pure algal cultures, for example, remain high in spite of innovations in production that make small batch enhancement of juvenile diets cost-effective. Better management of plankton and microbial communities in culture systems is increasingly understood to be a key requirement for maintaining high health stocks. So-called green-water culture, where water is conditioned in separate ponds stocked with finfish prior to use for shrimp culture, or finfish are stocked in cages within the shrimp pond, has been shown to inhibit the growth of disease-causing bacteria (Tendencia et al., 2006). This effect has been attributed to the proliferation of beneficial bacteria, fungi, and phytoplankton populations that develop as a consequence of selective feeding by the finfish and the subsequent excretion of nutrients, stimulating production.

Highly controlled systems, equivalent to "precision agriculture," contrast with conventional pond-based approaches in which feeding inefficiencies are compensated for through the use of lower-cost, less-sophisticated feed formulations. In such ponds, natural food webs themselves act as "environmental services," providing in situ treatment, provided densities and feeding levels do not exceed a relatively low carrying capacity. This continues to be the most cost-effective alternative in many, especially tropical, production systems dominated by herbivorous and omnivorous species. Such systems work particularly well where the major market is for individually small fish (<250 g). Technologies for efficient, semi-intensively managed pond systems are based on the combined use of fertilizers and pelleted feed, and typically depend on polycultures of several species. Floating, formulated diets allow efficient feeding of targeted species, as demonstrated by lower feed conversion factors, resulting in less waste remaining in the pond. A tendency for a reduced number of species in such feed-fertilized polycultures has emerged in areas of Asia with well-developed aquaculture (Edwards, 2012). This trend appears dependent on a competitive market for feeds formulated for a more valuable targeted species stocked together with "service species," such as the silver carp, to improve water quality by filter feeding plankton and carnivorous species to remove unwanted recruits. The persistence of polycultures, an intrinsic part of traditional low-input systems, demonstrates their continued relevance in modern commercial systems. In addition to their ecological function, they can deliver key benefits in terms of improving cash flow to pond operators and provide low-cost food for local consumption (Mamun et al., 2014). Both of these attributes contribute to the socioecological resilience of communities where aquaculture is located and further afield (Troell et al., 2014).

Polyculture can also be viewed at a broader scale, for example, using pond water and sediments to fertigate surrounding crops (Karim et al., 2011), which is a further example of aquaculture supporting food security and the indirect role ponds managed for fish production can play in foodscapes (Little et al., 2012).

5. Life Stage and Technological Choices: Many Opportunities for Juveniles, Fewer for Food Fish

The differential value of farmed organisms at different life stages may give rise to a range of technological options that make business sense. This is easily understood for feeds where investment in high-quality (and high-priced) juvenile feeds to achieve high survival and rapid early growth is often not sustained into a longer grow-out phase where a different cost–benefit relationship prevails. It is also often apparent for production systems; whereas parr to smolt Atlantic salmon production increasingly occurs in RAS, grow-out remains uneconomic in such high-cost systems (Box 1).

The separation of life stages—egg hatching, hatchling to fry, fry to fingerling—is often required because of the very different conditions (i.e., water depth, feed type, and delivery mechanism) required to optimize performance, but further differentiation may be desirable. The maintenance of stocking densities at levels that are close to carrying capacities remains a challenge in many systems. As intensification occurs, maintaining the optimal levels of feeding level, fish density, and water quality management becomes critical to ensuring profitability. A capacity for frequent but low-stress grading of stocks during juvenile rearing can be critical for reducing the size variation that can stimulate cannibalism. This occurs over a wide range of species, but is facilitated by good system design, whether shallow tank units or hapa cages.

Maintaining optimal fish densities in earthen ponds efficiently, especially larger or poorly designed units, can be a major challenge and probably explains a good deal of their variable performance. A cornerstone of the efficient production of tilapias in Chinese earth pond systems is the frequent thinning out of the stock by careful seine netting at intervals through the culture cycle. Such management optimizes the efficiency of pond use while maintaining individual fish growth. A similar situation exists in the management of *Pangasius* catfish in Bangladesh, where service fish, initially

Box 1 Emerging Technologies Enabling Land-Based Aquaculture in RAS and IMTA Systems

Effective water filtration to remove waste material and associated breakdown products is critical to developing land-based aquaculture systems. Filtration processes depend on passing the water through screens and media, and this necessitates the use of pumps. At higher stocking densities, aeration and oxygen injection may be required. Ozonation can be used to oxidize organic molecules and nitrite and to sterilize the process water. Ultraviolet lamps can also be used for process water sterilization and to reduce the risk of diseases being introduced in incoming water. A tendency for compounds, such as ammonia and nitrite, to accumulate in process water has demanded the development of biological filters to convert these compounds to less dangerous nitrate. Elevated concentrations of these substances, which also constitute plant nutrients, led to the development of aquaponics systems, where typically salad crops are grown with their roots in the fish culture water. Similar systems using a combination of predominantly aquatic species with different feeding habits to make the best use of available nutrients has been termed integrated multitrophic aquaculture (IMTA). Limitation to such land-based systems include the costs associated with their establishment and operation, notably electricity to power the pumps; risks associated with mechanical failures and disruption to electricity supplies; and the continued need to dispose of sludge and a proportion of the process water to avoid the accumulation of toxic compounds. Land-based culture systems can be financially viable, but often they have to compete with products from other systems with lower overheads. Financial returns can also be sensitive to modest changes in the efficiency of production and value of products being cultured.

"overstocked," are regularly removed throughout the growing period. This approach, where stock management through the growing cycle is essentially contracted out to netting teams, increases local employment benefits often to some of the poorest people. Compared to "all-in, all-out" stocking and harvest, improved productivity in terms of land and water use is also achieved.

A prerequisite in both cases is for ponds designed for efficient seining and the availability of a flexible, low-cost labor force able to work on such contractual terms.

6. Risk Management and Biosecurity for Animal and Environmental Health

Sick and dead fish contribute little to human food security, and ensuring high health and welfare of cultured seafood has been identified as a key issue to ensuring more productive and cost-effective aquaculture. However, the reality is that very high and inconsistent mortalities are the norm in many systems, and that if mortality occurs at an early stage, the overall economic consequences can be mitigated through stocking of larger numbers of juveniles—or following significant early mortality events, restocking. Preemptive harvest to minimize losses is a common strategy, especially where capacity for diagnosis and treatment is limited. In practice, more sophisticated aquaculture is based on stocking fish for final culture at larger, more robust sizes after what is often a series of juvenile stages for which mortalities may be high but unit investment low. The farming of fecund fish has therefore spurred a somewhat different attitude to stock mortality than for terrestrial livestock equivalents.

Managing high health stocks of fish requires knowledge of, and best management practices (BMP) to control, the key risk pathways for infection by pests, parasites, and pathogens (3Ps). The BMP will vary greatly in relationship to the openness of the culture systems with the wider environment and the capacity for controlling entry of the 3Ps hence the attraction of biosecure systems that can be maintained as protected or secure environments in isolation from natural systems. Such systems have independent water supplies from groundwater or sterilizable surface water sources, and are partially or completely enclosed to avoid transfer of viable fish, their juveniles, or 3Ps vectors. The high water efficiencies of RAS require only topping up of additional, treated water.

In poorer countries where aquaculture provides local food as well as income, there are often contradictions to ensuring optimal pond preparation and implementing other BMPs, with farmers' objectives of optimizing entry of wild aquatic animals, which are valued for their taste and value (Morales and Little, 2007).

In reality, most culture systems will remain open or reliant on using large amounts of water for the foreseeable future, and alternatives, such as ensuring high health of stocks prior to and during production, require a different emphasis. Part of this is ensuring robust stocks able to withstand certain levels of exposure to the 3Ps under normal culture conditions—to some extent an outcome of domestication and a passively selected trait. This can be enhanced through selective breeding (see the next section) and through ensuring optimal nutrition, water quality, and general husbandry that ensures low levels of stress, i.e., good system design and management (see the preceding section). Further improvements are likely through a better understanding of host parasite interactions, leading to more effective integrated pest management strategies. For example, approaches to better control Atlantic salmon sea lice in open-water cages include alternatives to conventional chemotherapeutants based on botanical products with more complex action pathways, development of effective vaccines, alteration of cage depth (based on an understanding of sea lice behavior), lasers and cleaner fish to physically reduce numbers in the environment or on the fish themselves, and release of sterile adults. Enhancing the immunological response of fish to infections of various types includes a range of approaches, from the use of probiotics to specific vaccines.

In general, understanding animal–environment interactions is proving critical to effective responses to health management. The pandemic of acute hepatopancreatic necrosis disease (AHPND) on white-legged shrimp in Asia points to the complexity of managing health of farmed aquatic animals. A technology-led approach based on frequent and high levels of disinfection within biosecure systems appears to have exacerbated the problem; such systems suffered greater losses than open landscape systems. The importance of understanding and managing sediment ecology has been identified as a key part of control (De Schryver et al., 2014).

A critical aspect of control of the 3Ps comes down to the soft-side of technology development and implementation—enhanced monitoring and surveillance on the farm, among concentrations of producers, and between countries (Bunting and Stentiford, 2014). This requires both improved institutional mechanisms and governance for each level that supports controls over transfers of fish and fish products, and those involved in their production and trading. It also has implications for how aquaculture is managed in the wider sphere—for example, the broader movement of goods and services that may facilitate the transfer of potentially invasive aquatic species (Havel et al., 2015) and the 3Ps. Effective control over ballast water in cargo ships and controlling imports of live fish are obvious issues. Faster and cheaper diagnosis of the 3Ps at the farm level is emerging as an essential tool—through rapid virus detection kits for shrimp at the pond side to the spatial mapping of pathogen spread between countries that could inform controls and reduce infection. Earlier diagnosis and sharing of intelligence, for example, could have reduced the economic impact of AHPND that spread over several months throughout Asia and now other shrimp-producing areas (Chamberlain and Lighner, 2013). Models for such systems are already developed for human health epidemics, with a key difference that competent authorities need to work more impartially and proactively, with commercial producers having no short-term economic incentive for sharing such information. Epidemiological approaches to understanding the spread and impacts of the 3Ps have proved useful for improving policy and should be used more widely (Stentiford and Thitamadee, 2014).

The close relationship between heath status and welfare of the farmed animal is now accepted and making rapid advances for some species and systems, often in advance of terrestrial livestock. The rapid development and industry take-up of the Freedom Food standard among Scottish salmon producers, an animal welfare-orientated standard in the United Kingdom, has been related to its market value to retailers and an economic advantage among producers and others in the value chain (SEAT, 2013).

7. Species or Strain Opportunities for Genetic Gain

Breeding more robust and productive animals that meet evolving consumer demands is becoming a key component of technology development for sustainable aquaculture. Until recently constrained by the problems of small founder stocks common in pioneer aquaculture, the most important species are generally now the focus of some form of selective improvement programs allied with conservation efforts for their wild forbears and environments they inhabit.

Establishing effective supply chains for juvenile fish is a key foundation step for the spread of aquaculture and has many implications for the type and cost of technologies best used. Improving the quality of juveniles, both in terms of their genetic and non-genetic characteristics, should result in improved productivity for the grower and, ultimately, lower prices for the consumer. Access to improved strains and technologies, such as techniques required to produce monosex[5] seed, is however unlikely to be scale-neutral and could exacerbate inequalities and potentially exclude poorer producers. Assessing the availability and affordability of hatchery technologies is important if benefits are to be shared through aquaculture value chains; these are issues common to the supply of other agricultural inputs, whether hybrid maize or improved livestock strains. Some key species with an important role in supporting food security in LMIC, such as tilapia, were introduced because of the simplicity of their seed supply—they can be produced simply as a by-product of the food fish production (Little and Hulata, 2000)—but improving quality further will generally be based on the introduction of some level of specialized seed producers. But how simple can systems that deliver improved seed be, and how can poorer members of aquaculture value chains retain or enhance their benefits? Use of improved strains and monosex, rather than mixed sex, tilapia did not deliver better outcomes in one comparative study in Vietnam (Dan and Little, 2000), and the developmental impacts in Bangladesh for decentralized approaches to production were found to be far greater than centralized monosex hatcheries (Haque, 2008). Not only did a wide range of poor supply chain actors have employment benefits from production of mixed sex tilapia in simply modified irrigated rice fields, but also the benefits to food farmers and consumers of stocking them in ponds because of their large size and local availability had major impacts. Critical among these benefits was the food supply smoothing during periods of low incomes and limited wild fish availability (Haque et al., 2010). An interesting outcome was the understanding that small fish, both stocked and wild, from such systems were critical to poorer peoples' food security, and this is believed to be a much broader phenomenon (Little et al., 2012). The consumption of whole small fish is generally less wasteful and nutritionally superior to consuming the flesh of larger fish (Thilsted, 2012).

This suggests that the objectives for "improving" the genetic status of fish in LMIC to benefit poorer peoples' food security may need revisiting. Rather than the typical focus on fast growth rate to produce individually large fish, geneticists should focus on breeding more robust fish that can produce higher yields of smaller fish, of which a greater proportion of harvestable yield can be consumed rather than plate wasted. Additionally, the nutritional value of species and size of fish should be considered in prioritizing their production. Life-cycle analyses suggest that culturing species and strains requiring limited or no given feed is a key issue for increasing the global catch in the most sustainable way (Henriksson et al., in press).

From a food security perspective, the challenge for 2050 will, given current trends, be the rise in purchasing power and the likely move to diets based on more meat and higher-value marine carnivorous species. The ability of breeders to develop strains of such marine carnivores capable of consuming a more terrestrial-based diet will be a priority, and progress has already been made in this direction for the rainbow trout (Overturf et al., 2013). In tandem, there will also be incentives for plant breeders to develop varieties of soy and other plants yielding more digestible and utilizable

[5] Monosex tilapia seed is produced using hormone supplementation of larval diets, and stocks arising from such technological interventions are desirable as they improve the proportion of larger, more valuable fish at harvest.

protein and micronutrients. Undoubtedly, research and development toward better processing of such ingredients (cold enzyme technology, etc.) will be critical parts of this emerging toolbox and of even greater significance for further improving terrestrial livestock efficiencies.

A key difference between terrestrial and aquatic livestock farming is the risks associated with genetic modification (used in the broadest sense) and its potential impacts on biodiversity. Whereas selective breeding has occurred largely in the absence of significant wild stocks for terrestrial livestock, it has been practiced with the same species in the same environments for most aquaculture. Maintaining the integrity of wild stocks is a key concern, particularly in areas where indigenous species are being cultured and farmed stocks are being modified through selective breeding. A question is the extent to which attempts at in situ conservation can be effective given the inevitable contamination of indigenous stocks. Implementing standards that reduce risks of escapes of "improved" germplasm or, through zonation, restrict aquaculture to less sensitive locations might be part of a strategy (Mattson et al., 2005). Consideration of landscape approaches based on undeveloped or less developed strains being produced in unimproved/naturalized farming habitats is another approach (Amilhat et al., 2010). This is directly analogous to the parallel trends obvious in livestock production, where intensive feedlot production based on improved breeds occurs alongside increasing demand for traditional breeds farmed under more natural conditions (Walton et al., 2015).

8. Toward Specialization or Integration: Aquaculture at a Crossroads

The incentives for further technological development appear relentless, given the challenge of meeting growth in demand for farmed aquatic products, but the nature of the technology adopted and dependence on external inputs required is likely to remain variable (Table 2). The use of high-precision food production in which high yields are obtained at improved levels of efficiency through optimizing management of the culture environment and use of external inputs, is already well advanced, and the yield gap between such systems and current economic enterprises well documented. For example, white leg shrimp can be produced at levels approaching 100 Mt/ha in well-designed and managed recirculating aquaculture systems, whereas pond-based commercial systems struggle to produce 10% of these yields, and for many smallholders using more extensive, resource-limited systems, much less than 10% of this level. Such high-precision systems have high capital and management requirements but can be located close to markets and, through exclusion of the 3Ps, control of water quality, and retention of wastes, can be highly productive without causing significant local environmental impacts.

Self-contained RAS systems (see Box 1) can reduce direct eutrophication impacts, but circulating the water within the system with mechanical pumps can result in a high greenhouse warming potential. Assessment of impacts across product value chains should adopt a life-cycle analysis approach to evaluate possible adverse outcomes

Table 2 Characterization of Aquaculture Systems Based on Reliance on Technology and Degree of Dependence on Externalizing Inputs

Aquaculture System	Reliance on Technology	Dependence on External Inputs
Rain-fed earthen fishponds	+	++
Semi-intensive fishponds (with on-demand water exchange)	++	+++
Raceways for salmonids	+++	++++
Intensive fish and shrimp ponds	+++	+++++
Land-based IMTA	++++	+++
Cage-based production (non-exposed sites)	+++	++++
Cage-based production (exposed sites)	++	+++
RAS	++++	++++
Super-intensive RAS	+++++	+++++
Marine finfish hatcheries	+++++	+++
Culture-based fisheries	+	+
Ranching	+	+
Extensive production in ponds and wetlands	+	+
On-bottom shellfish and plant production	+	+
Semi-intensive Integrated Agriculture-Aquaculture (IAA) systems	++	+/++
Fish-rice, prawn-fish-rice, and shrimp-rice	++	+/++
Shrimp-mangrove, fish-mangrove, and crab-mangrove	++	++/+++
Raft-based shellfish and plant production	++	+
Offshore long-lines for shellfish	+++	+
Organic production with own feed formulation	+++	++
Algae and rotifer culture	+++	+
Cage-based and open-water IMTA	+++	++
Aquaponics	++++	+++

Note: Reliance on technology (+) very low, (++) low, (+++) moderate, (++++) high, (+++++) very high; dependence on external inputs (+) very low, (++) low, (+++) moderate, (++++) high, (+++++) very high.

using a range of impact categories. Greater levels of dependence on external inputs associated with selected production systems (Table 2) can indicate that they are more vulnerable to changing access to and costs of inputs and that additional safeguards may be necessary to minimize risks to producers. Even when it is possible to culture some species, such as filter feeding bivalves with minimal technological inputs and consequently dependence on externalizing technology, it may be necessary to depurate[6] the animals prior to sale, thus incurring comparatively high energy costs to ensure their safe consumption (Newton, 2014).

Different efforts to reduce the environmental impacts of aquaculture intensification are likely to be required in line with an emerging dichotomy of aquaculture

[6] Depuration involves holding animals in clean water that is sterilized with UV light, for example, for a number of days to ensure that any pathogens are eliminated and do not pose a risk to consumers.

development. Integration and harmonization of production systems with other food production or with other key purposes, such as amenity or conservation uses of water, are already in evidence in LMIC and developed economies alike. In contrast, further isolation and specialization to optimize productivity is also evident in aquaculture value chains. Both trends are readily apparent in other forms of food production, and as market segmentation increases, demand for products from both is likely to increase.

Life-cycle analyses give some insights into how these two approaches compare in terms of global environmental impact; in general, more intensive and technology-supported production has lower per unit impact than lower output systems, often described as more ecologically benign. These findings are closely aligned with the perception that future food security can be assured through ecologically aligned technical progress (Bindraban and Rabbinge, 2012). However, fundamental to such assessments is the issue of defining boundaries. A recent study has indicated that integrated pig fish systems in China have, overall, lower impacts than when managed as separate monocultures (Zhang, 2014). Linking food systems through processing wastes can also reduce overall impacts; *Pangasius* catfish production in the Mekong Delta, Vietnam supports a significant volume of pork production through the use of reclaimed processing by-products as feed ingredients (Newton et al., 2014). A key emerging outcome is that aquaculture, through its reliance on water, nutrients, and land, will need to be competitive within the overall food systems on which the global population depends. As an outcome of this, some forms of aquaculture are likely to retain or further develop strong links with other societal activities. In parallel, specialization of knowledge and investment in increasingly globalized market economies will stimulate complete separation of some parts of the value chain—for example, genetic improvement and health products.

A key requirement for maintaining resilient aquatic food security is that innovation gains are accessible by stakeholders (e.g., by way of education, knowledge exchange programs, training, continuing professional development [CPD]) throughout farmed seafood value chains. Strong linkages with other users of the shared natural resource base, at both local and global scales, are required.

9. Future Trends

Enhanced technology can support sustainable intensification of aquaculture production, increasing the amount or value of aquaculture products while using resources more efficiently and mitigating environmental impacts. The value of prospective aquaculture technologies in a given context reflects the additional benefit delivered within the constraints imposed by the water, nutrient, and energy nexus.

The application of technology to any specific aquaculture development context has often been constrained by the competitive position of fish in markets compared to other animal source foods; using simpler systems has often produced fish more competitively, and this explains why ponds and cages systems dominate global production. Pressures to intensify are stimulating change in these conventional systems, however—for example, toward pond designs that maintain water quality using less energy. Such intensive ponds reportedly achieve yields threefold that possible in conventional ponds in China (Cremer et al., 2014). Investment in designs and scale of cage technologies make them suitable for more exposed sites with high water quality and reduce conflicts with other users. Such technological changes are not scale neutral and are likely to favor larger producers and corporate interests rather than smallholders. Such technologies will inevitably deliver greater labor efficiency, reducing employment on the farm but with potential gains elsewhere in the value chain. Such transformations have generally been associated with improved food security, but these have tended to occur in geographically limited clusters. Access to improved strains, feeds, and other inputs also remain far from universal, and their equitable distribution is often subject to market failure in LMIC.

Aquaculture as part of the broader food system and dependence on limiting resources is increasingly being regulated for its food safety and environmental and social credentials. Improved record keeping and traceability systems are therefore becoming a critical requirement.

Aquaculture systems have been typically defined by their technology characteristics and dependence on external resources. Increasingly, aquaculture is now being assessed from a broader systems perspective that frames the activity in terms of its overall resource use and the societal outcomes that result. Supporting further development that aspires to ensuring food security will require technology development to be aligned with broader innovation (Klerkx et al., 2012).

References

Amilhat, E., Lorenzen, K., Morales, E.J., Little, D.C., 2010. Fisheries production in Southeast Asian farmer managed aquatic systems (FMAS) II. Management impacts on diversity and productivity of aquatic resources. Aquaculture 298, 57–63.

Beck, M.B., Villaorel Walker, 2013. On water security, sustainability, and the water-food-energy-climate nexus. Frontiers of Environmental Science & Engineering 7 (5), 626–639.

Belton, B., Ahmed, N., Murshed-e-Jahan, K., 2014. Aquaculture, Employment, Poverty, Food Security and Well-being in Bangladesh: A Comparative Study. Penang, Malaysia: CGIAR Research Program on Aquatic Agricultural Systems. Program Report: AAS-2014-39.

Béne, C., Arthur, R., Norbury, H., Allison, E.H., Beveridge, M., Bush, S., Campling, L., Leschen, W., Little, D., Squires, D., Thilsted, S., Troell, M., Williams, M. Contribution of fisheries and aquaculture to food security and poverty reduction: assessing the current evidence. World Development, in press.

Bergheim, A., Brinker, A., 2003. Effluent treatment for flow through systems and European environmental regulations. Aquacultural Engineering 27, 61–77.

Bindraban, P.S., Rabbinge, R., 2012. Megatrends in agriculture – views for discontinuities in past and future developments. Global Food Security 1, 99–105.

Bostock, J., McAndrew, B., Richards, R., Jauncey, K., Telfer, T., Lorenzen, K., Little, D., Ross, L., Handisyde, N., Gatward, I., Corner, R., 2010. Aquaculture: global status and trends. Philosophical Transactions of the Royal Society B: Biological Sciences 365 (1554), 2897–2912.

Bunting, S., Stentiford, G., 2014. Regulatory regimes and disease controls (strategies, frameworks and methods to ensure robust and transparent sustainability). In: James, M. (Ed.), United Kingdom – South East Asia Aquaculture Workshops – Fish & Shellfish | Health & Nutrition. A Report Prepared for the UK Science & Innovation Network. MASTS, Scotland, pp. 16–18.

Bunting, S.W., Mishra, R., Smith, K.G., Ray, D., 2015. Evaluating sustainable intensification and diversification options for agriculture-based livelihoods within an aquatic biodiversity conservation context in Buxa, West Bengal, India. International Journal of Agricultural Sustainability 13(14), 275–294. http://dx.doi.org/10.1080/14735903.2014.966530.

Bush, S.R., Belton, Hall, D., Vandergeest, P., Murray, F.J., Ponte, S., Oosterveer, P., Islam, M.S., Mol, A.P.J., Hatanaka, M., Kruijssen, F., Ha, T.T.T., Little, D.C., Kusumawati, R., 2013. Certify sustainable aquaculture? Science 341, 1067–1068.

Brummett, R., 2013. Growing Aquaculture in Sustainable Ecosystems. Agricultural and Environmental Service Department Notes Issue 5, June 2013. The World Bank, Washington. 4 p.

Chamberlain, G., Lightner, D., 2013. EMS the Perfect Killer- a Webinar. As seen at Shrimp News International. As seen at http://www.shrimpnews.com/FreeReportsFolder/NewsReportsFolder/VietnamGAAwebinarEMS.html.

Cremer, M., Chappell, J., Zhang, J., Zhou, E., 2014. New Intensive Pond Aquaculture Technology Demonstrated in China. Global Aquaculture Advocate January/February, pp. 60–62.

Dan, N.C., Little, D.C., 2000. The culture performance of monosex and mixed sex new season and over-wintered fry in two strains on Nile tilapia (*Oreochromis niloticus*) in northern Vietnam. Aquaculture 184 (3–4), 221–231.

De Schryver, P., Defoirdt, T., Sorgeloos, P., 2014. Early mortality syndrome outbreaks: a microbial management issue in shrimp farming? PLoS Pathogens 10 (4), e1003919. http://dx.doi.org/10.1371/journal.ppat.1003919.

Edwards, P., 2012. Traditional Asian aquaculture. Ch. 34. In: New Technologies in Aquaculture. Woodhead Publishing, pp. 1029–1063.

FAO, 2002. Use of Fishmeal and Fish Oil in Aquafeeds: Further Thoughts on the Fishmeal Trap, by M.B. New & U.N. Wijkström. FAO Fisheries Circular No. 975, Rome. 61 pp.

FAO, 2014. The State of World Fisheries and Aquaculture 2014. Rome, 223 pp.

Fernquist, F., Ekelund, L., 2014. Credence and the effect of consumer liking of food. Food Quality and Preference 32, 340–353.

Garnett, T., Appleby, M.C., Balmford, A., Bateman, I.J., Benton, T.G., Bloomer, P., Godfray, H.C.J., 2013. Sustainable intensification in agriculture: premises and policies. Science 341 (6141), 33–34.

Haque, M.M., 2008. Decentralised Fish Seed Networks in Northwest Bangladesh: Impacts on Rural Livelihoods (Ph.D. dissertation). University of Stirling, 491 pp.

Haque, M.M., Little, D.C., Barman, B.K., Wahab, M.A., 2010. The adoption process of rice-field based juvenile fish production in Northwest Bangladesh: an understanding through quantitative and qualitative investigation. Journal of Agricultural Education and Extension 16 (2), 161–177.

Henriksson, P.J.G., Rico, A., Zhang, W., Nahid, S.K.A.A., Newton, R., Phan, L.T., Zhang, Z., Jaithiang, J., Dao, H.M., Phu, T.M., Little, D.C., Murray, F.J., Satapornvanit, K., Liu, L.N., Liu, Q., Haque, M.M., Kruijssen, F., de Snoo, G.R., Heijungs, R., Guinée, J.B. A Comparison of Asian aquaculture products using statistically supported LCA. Environmental Science & Technology, in press.

Havel, J.E., Kovalenko, K.E., Thomaz, S.M., Amalfitano, S., Kats, L.B., 2015. Aquatic invasive species: challenges for the future. Hydrobiologia 750, 147–170.

Karim, M., Little, D.C., Kabir, M.S., Verdegem, M.J.C., Telfer, T., Wahab, M.A., 2011. Enhancing benefits from polycultures including tilapia (*Oreochromis niloticus*) within integrated pond-dike systems: a participatory trial with households of varying socio-economic level in rural and peri-urban areas of Bangladesh. Aquaculture 314, 225–235.

Karapanagiotidis, I.T., Bell, M.V., Little, D.C., Yakupitiyage, A., Rakshit, S.K., 2006. Polyunsaturated fatty acid content of wild and farmed tilapias: effect of aquaculture practices and implications for human nutrition. Journal of Agricultural and Food Chemistry 54, 4304–4310.

Klerkx, L., Schut, M., Leeuwis, C., Kilelu, C., 2012. Advances in knowledge brokering in the agricultural sector: towards innovation system facilitation. IDS Bulletin 43 (5), 53–60.

Little, D.C., Hulata, G., 2000. Strategies for tilapia seed production. In: Tilapias: Biology and Exploitation. Fish and Fisheries Series, vol. 25. Kluwer Academic Publishers, Dordrecht, pp. 267–326.

Little, D.C., Barman, B.K., Belton, B., Beveridge, M.C., Bush, S.J., Dabaddie, L., Demaine, H., Edwards, P., Haque, M.M., Kibria, G., Morales, E., Murray, F.J., Leschen, W., Nandeesha, M.C., Sukadi, F., 2012. Alleviating poverty through aquaculture: progress, opportunities and improvements FAO/NACA. In: Subasinghe, R.P., Arthur, J.R., Bartley, D.M., De Silva, S.S., Halwart, M., Hishamunda, N., Mohan, C.V., Sorgeloos, P. (Eds.), Farming the Waters for People and Food. Proceedings of the Global Conference on Aquaculture 2010, Phuket, Thailand. September 22–25, 2010, FAO, Rome and NACA, Bangkok, pp. 719–783. 896 pp.

Little, D.C., Murray, F.J., Azim, E., Leschen, W., Boyd, K., Watterson, A., Young, J.A., 2008. Options for producing a warm-water fish in the UK: limits to "Green Growth"? Trends in Food Science and Technology 19, 255–264.

Mamun, A.A., Little, D.C., Bell, G., Murray, F.J., Kamaruddin, R., 2014. Salinity Impacts of Smallholder Shrimp and Prawn Aquaculture and Food Security in SW Bangladesh. Paper given at the European Aquaculture Symposium, San Sebastian, Ocotober 5–10, 2014.

Mattson, N.S., Somboon, N.S., Nguyen, M.T., Ouk, V., 2005. Aquaculture in the Mekong basin: alien or indigenous species? In: Bartley, D.M., Bhujel, R.C., Funge-Smith, S., Olin, P.G., Phillips, M.J. (Eds.), International Mechanisms for the Control and Responsible Use of Alien Species in Aquatic Ecosystems. FAO, Rome, pp. 127–131. Report of an Ad Hoc Expert Consultation, Xishuangbanna, People's Republic of China, August 27–30, 2003.

Maxwell, S., 1996. Food security: a post-modern perspective. Food Policy 21 (2), 155–170.

Morales, E.J., Little, D.C., 2007. Self-recruiting species (SRS) form farmer-managed aquatic systems-the contribution of non-stocked species to household livelihoods. FAO Fisheries Technical Paper No. 501. In: Bondad-Reantaso, M.G. (Ed.), Assessment of Freshwater Fish Seed Resources for Sustainable Aquaculture. FAO, Rome, pp. 603–616. 628 p.

Muir, J., 2005. Managing to harvest? Perspectives on the potential of aquaculture. Philosophical Transactions of the Royal Society of London. Series B, Biological Sciences 360 (1453), 191–218. Available at: http://www.pubmedcentral.nih.gov/articlerender.fcgi?artid=1636102&tool=pmcentrez&rendertype=abstract (accessed 11.15.12).

Naylor, R.L., Goldburg, R., Primavera, J.H., Kautsky, N., Beveridge, M.C.M., Clay, J., Folke, C., Lubchenco, J., Mooney, H., Troell, M., June 29, 2000. Effect of aquaculture on world fish supplies. Nature 405, 1017–1024. http://dx.doi.org/10.1038/35016500.

Newton, R., 2014. Environmental Sustainability and Value Addition of By-products Form Aquaculture Using Life Cycle Assessment (Ph.D. dissertation). University of Stirling, 447 pp.

Newton, R., Telfer, T., Little, D.C., 2014. Perspectives on the utilisation of aquaculture co-product in Europe and Asia: prospects for value addition and improved resource efficiency. Critical Reviews in Food Science and Nutrition 54 (4), 495–510.

Overturf, K., Barrow, F.T., Hardy, R.W., 2013. Effect and interaction of rainbow trout strain (*Oncoryhnchus mykiss*) on diet type and nutrient retention. Aquaculture Research 44, 604–611.

Pham, L.T., 2014. Sustainable Development of Export-orientated Farmed Seafood in the Mekong Delta, Vietnam (Ph.D. dissertation). University of Stirling, 378 pp.

Phuong, N.T., Oanh, D.T.H., 2010. Striped Catfish Aquaculture in Vietnam: A Decade of Unprecedented Development in Success Stories in Asian Aquaculture EDS. DeSilva, SS; Davy, FB 131-147. Springer, USA.

Roos, N., Wahab, M.A., Chamnan, C., Thilsted, S.H., 2007. The role of fish in food-based strategies to combat vitamin A and mineral deficiencies in developing countries. The Journal of Nutrition Symposium: Food-Based Approaches to Combating Micronutrient Deficiencies in Children of Developing Countries, 1106–1109.

SEAT, 2013. For the Good of the Fish. SEAT and Freedom Food. Available at: http://seatglobal.eu/2013/04/seat-the-european-perspective/#.Vfsn2d9Viko (accessed 17.09.10).

Stentiford, G.D., Thitamadee, S., 2014. Named pathogen/disease – special session – current work programmes on AHPND/EMS in Asia. In: James, M. (Ed.), United Kingdom – South East Asia Aquaculture Workshops – Fish & Shellfish I Health & Nutrition. A Report Prepared for the UK Science & Innovation Network. MASTS, Scotland, pp. 19–20.

Tendencia, E.A., dela Pena, M.R., Choresca, C.H., 2006. Presence of snapper, seabass, and siganid inhibits growth of luminous bacteria in a simulated shrimp culture system. Aquaculture 260, 54–60.

Thilsted, S.H., 2012. Improved management, increased culture and consumption of small fish species can improve diets of rural poor. In: Sustainable Diets and Biodiversity: Directions and Solutions for Policy, Research and Action. FAO and Bioversity International, Rome, pp. 176–181. http://www.fao.org/food/human-nutrition/sustainable-diets-and-biodiversity/en.

Troell, M., Naylor, R.L., Metian, M., Beveridge, M., Tyedmers, P., Folke, C., Arrowf, K.J., Barrett, S., Crépin, A.S., Ehrlich, P.R., Gren, A., Kautsky, N., Levinj, S.A., Nyborg, K., Österblom, H., Polasky, S., Scheffer, M., Walker, B.H., Xepapadeas, T., Zeeuw, A.D., 2014. Does aquaculture add resilience to the global food system? Proceedings of the National Academy of Sciences. www.pnas.org/cgi/doi/10.1073/pnas.1404067111.

Wahab, M.A., Kadir, A., Milstein, A., Kunda, M., 2011. Manipulation of species combination for enhancing fish production in polyculture systems involving major carps and small indigenous fish species. Aquaculture 321, 289–297.

Walton, M.E.M., Vilas, C., Coccia, S., Green, A.J., Cañavate, J.P., Prieto, A., van Bergeijk, S.A., Medialdea, J.M., Kennedy, H., King, J., Le Vay, L., 2015. The effect of water management on extensive aquaculture food webs in the reconstructed wetlands of the Doñana Natural Park, Southern Spain. Aquaculture 448, 451–463.

Zhang, W., 2014. Sustaining Export Orientated Value Chains of Farmed Seafood in China (Ph.D. dissertation). University of Stirling, 420 pp.

Zhou, Y., Du, S., Su, C., Zhang, B., Wang, H., Popkin, B.M., 2015. The food retail revolution in China and its association with diet and health. Food Policy 55, 92–100.

Part Two

Environmental and Policy Issues Affecting Food Security

Water Management

Helen Fyles, Chandra Madramootoo
Faculty of Agriculture and Environmental Sciences, Macdonald Campus, McGill University, Ste-Anne-de-Bellevue, QC, Canada

6

1. Introduction

Agriculture accounts for 70% of global freshwater withdrawals worldwide and more than 90% of its consumptive use (Gleick et al., 2009; Postel et al., 1996). Of the 1.5 billion hectares of cultivated land on earth, roughly 300 million hectares are equipped for irrigation and provide almost 40% of the world's food (FAO, 2013; Siebert et al., 2005). The majority of the world's irrigated area (70%) is in Asia, where China and India have 54 and 62 million hectares, respectively (Frenken and Gillet, 2012). This irrigated land accounts for 70% and 50% of their food production, respectively, and is critical to maintaining their food security (World Bank, 2006). In Sub-Saharan Africa, 97% of the total cropland is rain-fed, and these farmers are exposed to unpredictable weather conditions. Doubling the irrigated area in Sub-Saharan Africa could increase irrigation's contribution to the food supply from only 5% now to an optimistic 11% by 2050 (World Bank, 2013).

Water resource management is critical to optimize food production. In wet humid regions and low-lying coastal regions, excess water must be removed by drainage systems to achieve crop production, and supplemental irrigation may occasionally be beneficial during short dry periods in the growing season. Semiarid and dry subhumid areas experience seasonal and spatially variable rainfall, and producers are reliant on irrigation and other water management techniques to sustain crop yields (Rockström et al., 2010). In arid zones, crop water needs exceed total rainfall and yields are dependent on irrigation, often in conjunction with drainage, throughout the growing season. Although there are many non-water factors that limit crop production—including nutrient-poor soils, lack of infrastructure, labor shortages, insecure land ownership, inadequate knowledge or availability of technologies, and poor access to credit and markets—the greatest risk for the farmer is whether there will be sufficient water during the growing season. Investments in water management practices, such as rainwater storage, soil management to optimize water use, irrigation, and drainage, can reduce this risk (Rockström et al., 2010).

Provision of adequate water resources to meet human needs is already a severe problem in many parts of the world (Figure 1). Approximately 1.2 billion people worldwide live in river basins with absolute water scarcity where water resource development has exceeded sustainable limits. A further 1.6 billion suffer from inadequate access to water because of a lack of infrastructure or the human and financial capital to tap the available resources (Molden, 2007). In the populous and highly developed basins of South Asia and China, water use for irrigation as a share of total renewable

Emerging Technologies for Promoting Food Security. http://dx.doi.org/10.1016/B978-1-78242-335-5.00006-8

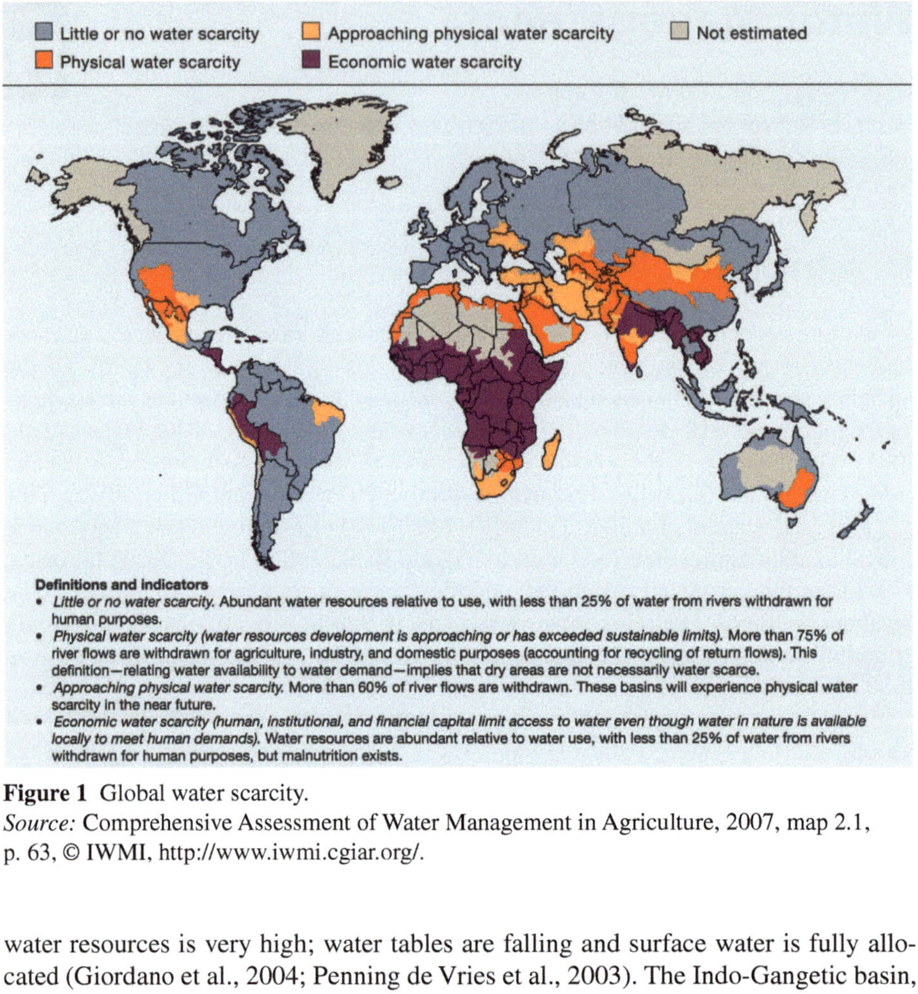

Figure 1 Global water scarcity.
Source: Comprehensive Assessment of Water Management in Agriculture, 2007, map 2.1, p. 63, © IWMI, http://www.iwmi.cgiar.org/.

water resources is very high; water tables are falling and surface water is fully allocated (Giordano et al., 2004; Penning de Vries et al., 2003). The Indo-Gangetic basin, for example, is home to a rapidly growing population of almost a billion people and produces most of the wheat and half of the rice grown in India and Pakistan (Sharma et al., 2010). A strong dependence on nonrenewable groundwater and glacial melt, particularly in the Indus basin, make food security in this region very vulnerable in the future. In many parts of North America and Australia, surface water use for irrigation is already capped and groundwater use is restricted to prevent long-term water shortages.

With growing global water scarcity, rising demand for water from nonagricultural sectors, uncertainties in water supply brought about by climate change, a current reliance on nonrenewable groundwater for about 20% of irrigation needs worldwide (Wada et al., 2012), coupled with a growing population requiring more food, it is evident that the agricultural sector must produce more food with less water. In addition, environmental degradation due to poor water and fertilizer management and health issues associated with increasing use of wastewater or contaminated water for irrigation must be dealt with.

2. Practices to Raise Water Use Efficiency (WUE)

Irrigation produces yields that are generally two to three times those of rain-fed agriculture, stabilizes crop production, raises farm incomes, improves crop quality, and allows for diversification in farm production, including the introduction of new crops that are not generally viable under dryland farming. Irrigation plays a critical role in food security, particularly as rising demand for food and climate change place increasing demands on producers. Irrigation also has the potential to exacerbate local water shortages, contaminate soil and water resources, and draw down nonrenewable water resources. Innovative irrigation technology, water storage, and scheduling are all tools that can be used to intensify food production on irrigated land. In southern Alberta, Canada, improved technology and management have resulted in higher crop water productivity. Barley production increased from 2.4 to 6.5 t/ha between the 1950s and 2000, while irrigation applications dropped from 900 to 380 mm and today the irrigation industry uses 30% less water to grow more crops than 25 years ago (Paterson, 2011).

One of the limitations impeding effective water management is irrigation supply technology. In many irrigation schemes, water is proportionally distributed to producers, as it is the easiest way to allocate water, even though this may create areas of shortage and areas of oversupply. To address this problem, provision of irrigation water is moving away from supply management to more intensive demand management, which is controlled by crop needs and holds excess water in storage for other uses or for times of water scarcity. Technology and infrastructure play a role in this transformation. In Mexico, the Yaqui Irrigation District practices rotational water distribution using adjustable water control structures and maintaining flexible water supplies for farmers according to their needs (Birendra et al., 2011). Soft tools for improving the operation, maintenance, monitoring, and evaluation of irrigation systems are increasingly available. Canal operation simulation models can be useful for the comparison of various design alternatives for irrigation canal systems, the development and tuning of operational strategies and automatic control algorithms, and for operation and training (Goussard, 2000). Modern tools like MaSSCOTe (mapping system and services for canal operation techniques) from the Food and Agriculture Organization of the United Nations (FAO); benchmarking of irrigation schemes jointly developed by the FAO, the International Water Management Institute, and the International Commission on Irrigation and Drainage; Internet-based weather forecasting and irrigation scheduling; and the FAO's Aqua Crop and systems dynamic modeling have been successfully tried in improving the performance of irrigation schemes (Renault et al., 2007; Vaez Tehrani et al., 2013). In addition, remote sensing and geographic information systems are being used for assessment of crop evapotranspiration and stress, irrigation potential, extent of waterlogging and soil salinity, and assessment of crop productivity, all of which can be used to increase crop water use efficiency (WUE) and reduce soil and water degradation.

2.1 Off-Farm Technologies

Water is conveyed from source (storage reservoir, river, or well) to field over varying distances to deliver irrigation water at a desired flow rate and at predetermined times and locations. Large amounts of water are often lost during conveyance, and technologies have been developed to improve the operation, regulation/control, and distribution of water over the command area. These include canal lining, upstream/downstream controls, supervisory control and data acquisition (SCADA),[1] total channel control (TCC), improved flow measuring and regulating devices, the use of telemetry, modern water delivery outlets, and en-route storage. In Australia, for example, modernization of infrastructure in the Goulburn-Murray Irrigation District with automated technology and repairs to outdated channels aims to increase WUE to 85% and recover long-term average annual water savings of 425 GL, with water savings shared between irrigators, the environment, and urban water users (DEPI, 2014).

2.2 On-Farm Technologies

Once water reaches the farm, it is applied to the crop by methods that can be broadly classified as gravity or surface methods, sprinkler, and micro-irrigation. Worldwide, the gravity system is the most dominant method, adopted on about 244 million ha (86% of total irrigated area), while the sprinkler-irrigated area is about 32 million ha (11%), and the micro-irrigated area about 8 million ha (3%) (Madramootoo and Gopalakrishnan, 2011).

Field application efficiencies tend to increase from gravity methods to sprinkler to micro-irrigation (Table 1), although there have been significant improvements in the application efficiency of surface methods: field channels are being replaced by piped conveyance systems; furrow and basin layouts are used on laser-graded fields; siphon tubes, gated pipes, and solar power-operated surge valves and cablegation regulate water flows into furrows (Madramootoo and Gopalakrishnan, 2011). As irrigation water availability dwindles, farmers in many arid and semiarid countries are

Table 1 Irrigation Efficiencies of Some Application Methods

Irrigation Method	Range[a] (%)
Gravity—Undeveloped	30
Gravity—Developed	40–80
Sprinkler—Hand-moved	60–85
Sprinkler—Wheel-moved	60–85
Sprinkler pivot (low pressure)	75–95
Micro-drip trickle	70–95
Low-energy precision application	75–98

[a]The range reflects possible variation due to factors such as upkeep, age, or nozzle type that can affect system application efficiencies.
Data from Howell (2003) and Schneider et al. (2000).

[1] SCADA refers to a system that collects data from various sensors at (remote) locations like the irrigation district control office or a factory or plant, and then sends the data to a central computer, which then manages and controls the data.

opting for high-efficiency micro-irrigation systems that provide a constant supply of water in the crop root zone and result in a higher crop yield, increased WUE, and lower energy consumption over conventional irrigation methods (Madramootoo and Morrison, 2013). In on-station trials in Niger, experiments comparing drip irrigation and irrigation with watering cans showed that drip irrigation helped achieve higher yields, higher returns to water, and higher returns to labor (Woltering et al., 2011). China and India are at the forefront of adopting micro-irrigation, covering about 20% of the world's irrigated area by these methods. Micro-irrigation can also be used to apply fertilizer in liquid form, reducing the application of excess nutrients and increasing crop yields.

Rice production uses roughly one-third of the world's freshwater (Bouman, 2009), and techniques such as an alternate wetting and drying (AWD) systems, broad ridges and furrows, and use of center pivots to irrigate rice with less water and raise crop yield are increasing. AWD, for example, can lower water use for irrigated rice by about 35% while improving yields and nutritional status (Price et al., 2013).

2.2.1 Irrigation Scheduling

Scheduling water applications to meet crop needs at critical growth stages can save water and increase crop growth and reduce runoff, energy consumption, soil water-logging, salinity, and soil erosion. Scheduling tools like tensiometers, wetting front detectors, remotely operated soil moisture sensors, and automatic weather stations are increasingly used in developed countries to allow weather data, soil moisture levels, and crop evapotranspiration to be included in water scheduling applications. Irrigation systems are generally designed based on average field and climatic conditions. Variable-rate irrigation (VRI) technology, using innovations in sensor technologies and wireless data communication networks, in conjunction with advances in Internet and mobile application technologies, is an important management tool for controlling the depth and timing of irrigation events to match irrigation water needs of individual areas within field. Although these tools are commercially available, they are unaffordable to small farmers. Even so, farmer training to better understand peak crop water demands and apply water when most needed can increase water and crop productivity.

With rising fuel costs and erratic power supplies in many developing countries, solar power is becoming increasingly effective and widely used as an energy source for irrigation systems. For example, in northern Benin, solar-powered drip irrigation significantly augmented both household income and nutritional intake, particularly during the dry season, and was cost-effective compared to alternative technologies (Burney et al., 2010).

2.2.2 Irrigation for Smallholders

An estimated 85% of the 525 million farms worldwide are operated by smallholders on plots measuring less than 2 ha (Nagayets, 2005). Improving water management on these farms to increase water productivity will make a major contribution to global food security. Irrigation carried out by smallholder farmers dominates the irrigation landscape in South Asia and is expanding rapidly in Sub-Saharan Africa (Xie et al., 2013). In India, more than half of all irrigation is drawn by pumps owned by smallholder farmers (Shah, 2009). In Sub-Saharan Africa, more than five million smallholder farmers are estimated to use small-scale technologies to irrigate more than one million hectares (de Fraiture and Giordano, 2014). To improve WUE of irrigation used by smallholders, inexpensive and simple irrigation systems have been developed to save labor and optimize the use of water for row crops on small plots (Palada et al., 2011). For example, Bucket Kits include a simple 20-l household bucket (often replaced with a polythene bag) attached to a pole at shoulder height and equipped with drip tapes to water a kitchen garden (Singh et al., 2009). Low-head drip irrigation technology using drip tape has proved useful in sloped and uneven fields, and has a high WUE. The International Crops Research Institute for the Semi-Arid Tropics developed African market gardens (AMGs), which use drip technology to grow high-value horticultural crops on a small scale. Training and institutional support for the farmer is critical for correct use of irrigation technology and successful returns on investments (Burney and Naylor, 2012).

2.2.3 Soil Management and Nutrient Availability

Soil management to improve soil moisture for crop use through reduced evaporation losses and to prevent erosion and nutrient and organic matter losses can be accomplished using resource-conserving technologies and practices (Table 3). The extent to which such interventions increase total food crop productivity was measured on 12.6 million farms (37 Mha or 3% of the cultivated area in developing countries) in 57 poor countries (Pretty et al., 2006). Average crop yield increase was 79%, and all crops showed WUE gains, with the highest improvement in rain-fed crops. Uptake of these techniques by farmers is affected by local conditions. For example, although conservation agriculture may provide long-term benefits to soil fertility and water holding capacity, severely degraded soils are unlikely to produce enough crop residues for use as mulch, and even where sufficient residues are produced, there may be competitive uses, such as fodder or fuel (Lahmar et al., 2012).

Development of management practices suitable to specific crops can also lead to improvements in plant water use. An example is a system of rice intensification (SRI), which involves early, quick, and healthy plant establishment; reduced plant density; improved soil conditions through enrichment with organic matter; and reduced and controlled water application (CIIFAD, 2014). Rice yields and water use efficiency from SRI are significantly higher than for rice grown using conventional methods (Zhao et al., 2010).

2.3 Rainwater Harvesting

Agriculture in semiarid tropical regions, particularly in Africa, relies predominantly on rainfall, which occurs in short, torrential downpours during an increasingly unpredictable wet season. Much of this water is lost as surface runoff, evaporation, or deep drainage, and rainwater use efficiency is low (35–45%) (Pathak et al., 2009).

Table 2 **Examples of Soil Management to Reduce Erosion and Improve Crop Yield**

Management Practice	Application
Terracing: reduce slope steepness or length to reduce water runoff and soil erosion	Grass strips: planted or left unploughed along contours. Grass used as fodder or mulch. Contour bunds: excavate a channel and create small ridge on downhill side used for draining excess runoff on steep slopes. Trash line: arrange previous crop residues or other dead plant material in lines across slope. Reliance on excess residues. Fanya Juu: dig trench along contour and throw soil upslope to form embankment that reduces slope length and erosion. Planting the bund can produce an alternate crop. Stone lines: stones arranged in lines across slope. Lines are permeable, slow runoff rate, filters water and spreads water over field. Bench: used in steep areas to achieve a level bench.
Micro-basins	Different shapes of small basins, surrounded by low earth bunds are formed to enable the runoff to infiltrate at the lowest point, where the plants are grown.
Trench farming: increase water infiltration, soil moisture, and fertility	Large beds created and large amounts of organic matter incorporated. Between beds are deep furrows that retain moisture and are used to grow plants for food and fodder during dry season.

Source: Mati (2005).

Table 3 **Resource Conservation Practices and Technologies**

Resource Conservation Practice	Objectives
Integrated pest management	Uses ecosystem resilience and diversity for pest, disease, and weed control, and seeks only to use pesticides when other options are ineffective.
Integrated nutrient management	Balances the need to fix nitrogen within farm systems with the need to import inorganic and organic sources of nutrients and to reduce nutrient losses through erosion control.
Conservation agriculture	Reduces the amount of tillage sometimes to zero so that soil can be conserved and available moisture used more efficiently; adds mulch to soil surface to reduce evaporation and add organic matter.
Agroforestry	Incorporates multifunctional trees into ag systems and collective management of nearby forest resources.
Aquaculture	Incorporates fish, shrimps, and other aquatic resources into farm systems, such as into irrigated rice fields and fish ponds, and so leads to increases in protein production.
Livestock Integration into farming systems	Use of dairy cattle and poultry, including using zero-grazing. Provides manure to fertilize fields.

Source: Pretty et al. (2006).

Rainwater harvesting (RWH) artificially collects rain to make it available for cropping or domestic purposes. The economic costs and benefits of the various micro and macro RWH techniques described below are highly influenced by the degree to which nutrient inputs are used to improve crop yields. For example, owing to the high cost of labor, transport, and material inputs for the installation of stone rows or grass strips in Burkina Faso, these measures were not cost effective without the addition of fertilizer (Zougmore et al., 2004).

Micro-catchment rainwater harvesting systems are designed to collect runoff from within the farm boundary. The runoff water is usually guided into a type of infiltration enhancement structure constructed in the field and used to grow plants. In Sub-Saharan Africa, commonly applied micro-catchment rainwater harvesting techniques include terracing or contouring, pitting, and micro-basins (Table 2). Some of the techniques may have different names in different regions and minor differences in design and use (Biazin et al., 2012), but all are designed to be controlled by the farmer, which makes these systems easy to adapt. Their suitability to a given region depends on biophysical factors (amount of rainfall, topography, crops grown, and soil type) and socioeconomic factors (availability of labor, access to credit).

In macro-catchment systems, the runoff is usually collected from external catchments, such as roofs, streets, or natural slopes, and diverted into well-designed storage structures, including cisterns, underground tanks, open ponds, or earthen embankments. According to Wisser et al. (2010), water storage in small reservoirs could increase global cereal production by 35–45% through supplemental irrigation. In South India, where there is little groundwater storage, groundwater abstraction is costly, and rivers are seasonal, irrigation from rain-fed reservoirs produces 4.2 million tons of rice (Palanisami et al., 2010). In Sub-Saharan Africa, cisterns are covered to reduce evaporation losses, and their walls are plastered to avoid seepage losses, although these improvements often make such systems too expensive for poor farmers (Biazin et al., 2012).

3. Wastewater Use

Recent estimates suggest that roughly one billion people consume agricultural products initially grown with wastewater, and that number is likely to increase as food needs rise and water becomes increasingly scarce (Drechsel and Evans, 2010). As cities expand, more urban and industrial wastewater, often rich in nutrients, will be generated and could be used in peri-urban market gardens and aquaculture. In Hanoi, Vietnam, for example, 80% of vegetables are irrigated with water mixed with wastewater (Molden, 2007). In the water-deficient countries of the Middle East and North Africa (MENA), there is an increasing need to use available water resources more efficiently, and exploiting the potential contribution of reclaimed water for food production may be worthwhile. However, contaminants in the wastewater can pose risks for human health and the environment, and food safety and other phytosanitary measures must be implemented (Qadir et al., 2010). According to Molden (2007), three main policy approaches for improving management of marginal-quality water are "reduce the amount of marginal-quality water generated, minimize the risks when it is used in agriculture, and minimize the risks when handling food grown with such water."

4. Excess Water Management

Floods can have far-reaching negative impacts on national economies through widespread damage to agricultural lands and infrastructure. In South and Southeast Asia and the Pacific, it is estimated that more than 13 million hectares of agricultural land are prone to floods (FAO, 2001). In some major rice-producing countries such as Thailand, Vietnam, India, Bangladesh, and Cambodia, floods are caused primarily by an accumulation of rainwater, glacial melt, and tidal movements. For example, in Bangladesh, increasing glacial melt from the Himalayas as a result of rising global temperatures swells the Ganges and Brahmaputra Rivers and their hundreds of tributaries, flooding 30–70% of the country each year as the water makes its way to the Bay of Bengal in the south (Ali, 2007). The coast is also vulnerable to flooding from rising sea levels.

Construction of embankments, dykes, sluices, etc., for flood protection, and the use of high-yielding crop varieties that mature and can be harvested prior to the onset of floods, can improve food production in these areas (FAO, 2001). Post-monsoon drainage is also a critical factor governing crop productivity in many areas.

4.1 Tidal Lowlands

Tidal lowlands in the humid tropics have a tremendous potential to significantly contribute to food production. In Thailand and Vietnam, major rice exporting countries, most of the rice produced comes from lowland areas, especially in the Chao Phraya (Thailand), Mekong and Red River (Vietnam) floodplains and deltas (Schultz, 2011). Due to tidal fluctuations, these lowlands may have the advantage that irrigation as well as drainage by gravity is possible, especially when the tidal movement takes place in the freshwater zone near the mouth of a river (Schultz, 2011). Agricultural schemes in such lowlands have a rational layout, with relatively large and rectangular parcels, making mechanization and increase in farm size possible. Adequate water management infrastructure includes canals (primary, secondary, tertiary) with movable gates to control water.

Although these lowland coastal areas are often very valuable for agricultural production, they are also vulnerable to flooding during the rainy season or from rising levels of seawater. In order to protect this food production, lowland areas must be managed accordingly. The government of Vietnam, for example, has recently adopted a policy to optimize food production in the Mekong Delta. The policy gives a high level of flood protection for the higher parts of the delta, and protection of the lower levels of the delta to allow two crops to be grown annually with inundation during peak annual flooding to enable the spreading of fertile natural sediments over the fields. There is no protection for the deepest parts of the delta, and these are used as storage areas (Schultz, 2011).

4.2 Agricultural Drainage

Over the last few decades, drainage has evolved from a single-purpose measure to drain waterlogged or even permanently submerged lands, primarily for agricultural purposes, to an essential element of integrated water management under multiple land uses (Schultz et al., 2007). A drainage system can create a well-drained arable land, prevent soil salinization and erosion, lower groundwater levels, and/or remove accumulated salts or toxic elements through the use of surface or subsurface drains (Shultz et al., 2007).

Roughly 60 million ha (or 20%) of the world's irrigated area is equipped with drainage systems, mostly in arid and semiarid zones, where the major role of drainage is to prevent irrigation-induced waterlogging and salinization of the soil (Zimmer et al., 2005). The majority of the world's drainage systems, about 120 million ha, are installed in rain-fed lands in the temperate and humid zones, where drainage is used for waterlogging control, salinization prevention, and flood protection.

While drainage is a critical part of agricultural production in many regions, loss of valuable wetlands; production of drainage waters with high levels of salt, nutrients, and pesticides; altered river flows; and increased sediment loads in surface channels are all environmental consequences that need to be addressed (Schultz et al., 2007; Schottler et al., 2014).

Agricultural reuse of saline drainage water for salt-tolerant forages, biofuel crops, or brackish water-based fisheries is possible where water is managed at a regional scale (Qadir et al., 2007). The final discharge of small quantities of saline water may be disposed of in evaporation ponds or water treatment plants. Controlled drainage can be used to increase crop nutrient uptake (Wesström et al., 2014) and also for subirrigation during dry periods.

5. Environmental Issues

Negative environmental impacts associated with irrigation include depletion of water resources due to overuse, salinization, and contamination of soil and water from poor irrigation practices and inadequate fertilizer management; removal of water from natural aquatic systems, reducing biodiversity; and diversion of water via dams and irrigation canals, which can increase breeding sites for mosquitos (WHO, 2003).

Irrigation taken from groundwater is roughly 70% of total groundwater abstractions worldwide (WWAP, 2012). Depletion of groundwater more than doubled between 1960–2000 and 2000–2009, and the fraction of global groundwater abstractions from nonrenewable groundwater has increased from approximately 11% to 15% (Döll et al., 2014). Recent levels of groundwater depletion in Northern India, the United States, China, and Pakistan—which have large areas of irrigated cropland—were 43, 23, 6, and 4 km^3/yr. Aquifer overpumping can lead to land subsidence, saltwater intrusion, and sea level rise, as well as compromise long-term food production.

Agriculture can be a major source of nonpoint pollution and water quality impairment due to erosion from agricultural lands and excess fertilizer, manure, and pesticide applications. Chemicals and sediments enter waterways through surface runoff, seepage, or drainage. Mismanagement of soil inputs coupled with poor irrigation technologies and timing compound the problem.

More efficient use of water for food production, particularly in water-scarce and vulnerable regions, use of best management practices to improve soil and nutrient management, and effective use of wastewater will contribute to a lower agricultural demand for water, reduce nonpoint pollution, and ameliorate environmental impacts. Management of water resources at the basin or watershed level with input from all stakeholders is necessary to fully understand the implications of diverting water flows, preventing negative environmental impacts, and making informed decisions for the benefit of the whole community, depending on the water resource. Government environmental policies to reduce the negative impacts of water management strategies need to be clearly articulated.

6. Institutions and Governance

Expanding rainfall capture, groundwater use, surface water diversions, and artificial storage or drainage systems for agricultural purposes upstream will alter the quantity and quality of downstream flows, which may be supporting other producers, fishermen, industries (e.g., hydropower, mining, forestry), household users, and aquatic ecosystems. Managing water resources to avoid scarcity; prevent waste, pollution, and conflict; and satisfy water users is therefore a very complex problem that requires a long-term perspective. Appropriate and accountable institutions and governance at the community, watershed, and/or river basin levels may be necessary to prevent environmental degradation, improve social inequity, and optimize economic efficiency.

6.1 Participatory Approaches

Participatory approaches to water management that bring together water users who understand local water resources as well as physical and social circumstances can be more effective in resolving local water resource issues than regional and national government agencies (Bruns et al., 2005). However, many water use problems, such as pollution and economic development, evolve beyond the community and require regional or national governance. As a result, different groups or agencies of various sizes and power may hold different and overlapping rights over the same water resource. There is no single institution or set of regulations that will solve all water management issues, but rather a range of options that address the unique circumstances of the community, watershed, or basin in question. A few of these options are outlined next.

Within a landscape, there are multiple uses and purposes for water, each valued in different ways by different stakeholders, and trade-offs need to be reconciled (Sayer et al., 2013). Integrated water resource management (IWRM) can highlight such trade-offs through the use of tools, such as cost–benefit analysis, valuation of nonmarket services, assessments of risk and vulnerability, and models for estimated water flow requirements of wetlands, and help decision makers to develop strategies to optimize water use (Molden et al., 2010). IWRM emphasizes the ideas of integration, decentralization, participation, and economic and financial sustainability, and uses the basin as the unit for decision making. Although excellent in principle, implementation of IWRM is often difficult, due to donor or top-down management and politicization of the process. In addition, key water management problems are rarely at the scale of the water basin (Giordano and Shah, 2014). According to these authors, a focus on solutions to specific problems, which include an understanding of the physical, social, and political context, may be more successful at addressing water management issues than universal, water-centered approaches.

At watershed or smaller scales, participation approaches have gained momentum over the last decades (Carr et al., 2012). A project involving two watersheds in China implemented IWRM over three years and improved rainwater management, harvesting, soil, crop, and pest management options, as well as income-generating microenterprises for the community members (Wani et al., 2013). Significantly higher yields, greater crop diversification, and larger family incomes resulted.

In some regions, the irrigation sector suffers from weak institutional management, system inefficiencies, failure to recover operational costs, failure to innovate and integrate new technologies into the irrigation network, and a lack of adequate operation and maintenance and rehabilitation of infrastructure. Where water is not equitably and uniformly distributed, confidence in centrally operated irrigation systems often declines, and farmers withdraw water improperly and illegally and install individual tubewells to supplement canal water supplies. The latter has led to groundwater exploitation and lowering of water tables in many parts of the world.

At the level of an irrigation scheme, water user associations or local collective action initiatives that share the cost of infrastructure, promote fair water allocations, and address user conflicts can often be successful at managing local water supplies (Burney et al., 2013). Land and water management practices and consequently food production per unit area and irrigation supply are also found to be better in farmer-managed irrigation schemes in comparison to agency-managed schemes (Birenda et al., 2011). Whereas the former keep irrigation service fees within a user association, in the latter a large proportion of the fees may be sent to the government and not reinvested in the irrigation scheme.

In order to boost performance of large irrigation systems, improved communication is also critical between water user associations (WUAs) and field staff who represent the irrigation agency or bureaucracy and who do the day-to-day management of the irrigation systems (Suhardiman and Giordano, 2014). These authors give an example from Indonesia, where field-level irrigation staff held regular meetings with local WUAs to address water scheduling issues. Although agreements were not always put into practice perfectly (due to illegal water taking, internal struggles within the bureaucracy, elite capture, and lack of cooperation between WUAs), the meetings were a useful platform where both farmers (WUAs) and field staff were accountable for their actions and communicated their concerns and opinions.

WUAs are not always appropriate for managing and maintaining water infrastructure at the community level. In their study of small reservoirs in Sub-Saharan Africa, Venot et al. (2012) observed that traditional authorities often played a crucial role in conflict management, while government officials played a key role in construction, procurement, and agricultural extension. In this situation, introducing a WUA as the only legitimate decision making body may be counterproductive, particularly if irrigators are overrepresented in WUAs. Institutions and governance put in place to manage water resources must be specific to the local institutional and political context.

6.2 Water Pricing

The most common economic incentive used to motivate water users to conserve and use water efficiently is water pricing, although, taxes, subsidies, and quotas are also implemented (Binswanger-Mkhize et al., 2009). Prices, however, are generally far below full capital cost recovery in both developing and industrialized countries, and prices high enough to induce significant changes in water allocation (or recover capital costs) could severely reduce farm income. Administration costs are also an issue in some countries. India, for example, has an average farm size of about 1 ha, and administering water pricing would likely result in significant transaction costs compared to potential water market gains (Poddar et al., 2014). However, water pricing policies can improve efficiency and sustainability when combined with appropriate supporting policies.

Approaches to improving irrigation performance include public–private partnerships (PPP). Within an irrigation scheme, public involvement tends to diminish from upstream to downstream, with the regulation and control functions staying under public sector responsibility and the investment and operation, maintenance, and management (OMM) functions falling under private administration (World Bank, 2007). In an example of a PPP in the Yangtze Basin, China, the Hunan government contracted out OMM for dam and main canals to private water supply companies and OMM for branch and lateral canals to WUAs (World Bank, 2007). The result was an end to infrastructure degradation and control over asset maintenance funds by the private groups. For much of the world, however, private-sector irrigation development has been limited to groundwater development and, to a lesser extent, smaller commercial surface water systems growing high-value crops. For example, all groundwater irrigation in the United States and two-thirds or more of groundwater irrigation in Bangladesh, India, and Mexico is privately managed (Binswanger-Mkhize et al., 2009). Private concessions for the development and sale of irrigation water are feasible if there are good prospects for profitability.

Many challenges to managing water resources lie ahead. Rising use of groundwater for irrigation, mainly by private users, has resulted in serious depletion of water tables that will require regional governance efforts to solve. The current rapid growth of small-scale irrigation, particularly in Sub-Saharan Africa, is largely unregulated and unplanned, and can have undesirable environmental and social consequences (Giordano and de Fraiture, 2014). Conflict between smallholders over water use may arise with little opportunity for resolution, as there is often no organization to coordinate water use. Issues of gender inequity may also be exacerbated as women are underrepresented in the use and ownership of small-scale irrigation equipment (Giordano and de Fraiture, 2014). Water management cannot be seen as simply a technical issue but also as a social and political issue.

Management of water resources at any scale needs to be aware of lower-level processes in order to improve local interventions and higher-level processes to inform and coordinate administrative entities (Sayer et al., 2013). Improved capacity is also needed to coordinate water users across systems and sectors, bridging between irrigation, municipal, industrial, and environmental uses and users (Binswanger-Mkhize et al., 2009).

7. Conclusions

Many tools, practices, and systems of land and water management have been developed to improve crop yields, water productivity, energy efficiency, soil fertility, and local food security. However, often these improvements do not reach the farmers who would benefit from them, are looked at independently and without regard for the entire agro-ecosystem, or are inappropriate for the conditions where they will be used. Understanding the unique political, cultural, and geographic (including soil and water) context of a particular region is a prerequisite for managing water resources appropriately and for raising local food security.

Managing water use at local or larger scales to avoid depletion of nonrenewable water resources, provide producers with a reliable, affordable, and timely water supply, and maintain flows to natural ecosystems and other sectors without conflict is the greatest challenge facing the agriculture sector. Water management is only one of many tools to achieve food security. Others include agrochemicals, soil and land management, germplasm, crop processing and storage, on- and off-farm infrastructure, education, extension services, and credit. However, providing producers with the tools and knowledge to grow their crops at the highest level of water productivity possible is a critical part of the solution to food insecurity.

References

Ali, A.M.S., 2007. September 2004 flood event in south-western Bangladesh: a study of its nature, causes, and human perception and adjustments to a new hazard. Natural Hazards 40, 89–111.

Biazin, B., Sterk, G., Temesgen, M., Abdulkedir, A., Stroosnijder, L., 2012. Rainwater harvesting and management in rainfed agricultural systems in sub-Saharan Africa–a review. Physics and Chemistry of the Earth, Parts A/B/C 47, 139–151.

Binswanger-Mkhize, Hans P., Meinzen-Dick, Ruth, Ringler, Claudia, 2009. Policies, Rights, and Institutions for Sustainable Management of Land and Water Resources. SOLAW TR09 http://www.fao.org/fileadmin/templates/solaw/files/thematic_reports/TR_09_web.pdf.

Birendra, K.C., Schultz, B., Prasad, K., 2011. Water management to meet present and future food demand. Irrigation and Drainage 60, 348–359.

Bouman, B.A.M., 2009. How much water does rice use. Rice Today 8, 28–29.

Bruns, B.R., Ringler, C., Meinzen-Dick, R.S. (Eds.), 2005. Water Rights Reform: Lessons for Institutional Design. International Food Policy Research Institute, Washington, DC. http://www.ifpri.org/pubs/books/oc49/oc49.pdf.

Burney, A.Q., Naylor, R.L., Postel, S.L., 2013. The case for distributed irrigation as a development priority in sub-Saharan Africa. Proceedings of the National Academy of Science United States of America 110, 12513–12517.

Burney, J.A., Naylor, R.L., 2012. Smallholder irrigation as a poverty alleviation tool in sub-Saharan Africa. World Development 40, 100–123.

Burney, J., Woltering, L., Burke, M., Naylor, R., Pasternak, D., 2010. Solar-powered drip irrigation enhances food security in the Sudano-Sahel. Proceedings of the National Academy of Sciences of the United States of America 107, 1848–1853.

Carr, G., Blöschl, G., Loucks, D.P., 2012. Evaluating participation in water resource management: a review. Water Resources Research 48 (11).

Cornell International Institute for Food, Agriculture and Development (CIIFAD), 2014. SRI Methodologies. http://sri.ciifad.cornell.edu/aboutsri/methods/index.html.

Department of Environment and Primary Industries (DEPI), January 2014. GM-W Connections Project. http://www.depi.vic.gov.au/water/rural-water-and-irrigation/improving-irrigation-efficiency/g-mw-connections-project.

Döll, P., Schmied, H.M., Schuh, C., Portmann, F.T., Eicker, A., 2014. Global-scale assessment of groundwater depletion and related groundwater abstractions: combining hydrological modeling with information from well observations and GRACE satellites. Water Resources Research 50, 5698–5720.

Drechsel, P., Evans, A.E., 2010. Wastewater use in irrigated agriculture. Irrigation and Drainage Systems 24, 1–3.

FAO, 2013. AQUASTAT Database, FAO's Global Information System on Water and Agriculture. Food and Agriculture Organization of the United Nations (FAO) (website accessed on January 2014).

de Fraiture, C., Giordano, M., 2014. Small private irrigation: a thriving but overlooked sector. Agricultural Water Management 131, 167–174.

FAO, 2001. Report of the FAO Asia-Pacific Conference on early warning, prevention, preparedness, and management of disaster in food and agriculture. Changmai, Thailand. June 12-15, 2001. RAP Publication 2001/14.

Frenken, K., Gillet, V., November 2012. Irrigation Water Requirement and Water Withdrawal by Country. AQUASTAT, FAO. http://www.fao.org/nr/water/aquastat/water_use_agr/index.stm (September 2013).

Giordano, M., de Fraiture, C., 2014. Small private irrigation: enhancing benefits and managing trade-offs. Agricultural Water Management 131, 175–182.

Giordano, M., Shah, T., 2014. From IWRM back to integrated water resources management. International Journal of Water Resources Development 30 (3) (published online January 7, 2014).

Giordano, M., Zhu, Zhongping, Cai, Ximing, Hong, Shangqi, Zhang, Xuecheng, Xue, Yunpeng, 2004. Water Management in the Yellow River Basin. Background, Current Critical Issues and Future Research Needs. Comprehensive Assessment Research Report 3. Comprehensive Assessment Secretariat, Columbo, Sri Lanka.

Gleick, P.H., Cooley, H., Morikawa, M., 2009. In: Gleick, P.H. (Ed.), The World's Water 2008–2009: The Biennial Report on Freshwater Resources. Island Press, pp. 202–210.

Goussard, J., 2000. Canal Operation Simulation Models; Provisional Catalogue of Models Currently Available. ICID, New Delhi. 27 pp.

Howell, T.A., 2003. Irrigation Efficiency, in Encyclopedia of Water Science. Marcel Dekker, Inc., New York. 1076 pp.

Lahmar, R., Bationo, B.A., Dan Lamso, N., Guéro, Y., Tittonell, P., 2012. Tailoring conservation agriculture technologies to West Africa semi-arid zones: building on traditional local practices for soil restoration. Field Crops Research 132, 158–167.

Madramootoo, C.A., Morrison, J., 2013. Advances and challenges with micro-irrigation. Irrigation and Drainage 62 (3), 255–261.
Madramootoo, A., Gopalakrishnan, M., 2011. Innovations in irrigation for food production in Asia. In: Madramootoo, C.A. (Ed.), Water Management for Global Food Security. McGill Institute for Global Food Security, McGill University, Montreal, Quebec. 122 pp.
Mati, B.M., 2005. Overview of Water and Soil Nutrient Management under Small Holder Rainfed Agriculture in East Africa. Working Paper 105 International Water Management Institute (IWMI), Columbo, Sri Lanka. http://www.iwmi.cgiar.org/Publications/Working_Papers/working/WOR105.pdf.
Molden, D., Oweis, T., Steduto, P., Bindraban, P., Hanjra, M.A., Kijne, J., 2010. Improving agricultural water productivity: between optimism and caution. Agricultural Water Management 97, 528–535.
Molden, D., 2007. Comprehensive assessment of water management in agriculture. In: Water for Food, Water for Life: A Comprehensive Assessment of Water Management in Agriculture. International Water Management Institute, London: Earthscan, and Colombo.
Nagayets, O., June 2005. In: Small Farms: Current Status and Key Trends. Information Brief Prepared for the Future of Small Farms Research Workshop. Wye College, Kent, UK, pp. 26–29.
Palada, M., Bhattarai, S., Wu, D.L., Roberts, M., Bhattarai, M., Kimsan, R., Midmore, D., 2011. More Crop Per Drop: Using Simple Drip Irrigation Systems for Small-Scale Vegetable Production. AVRDC–The World Vegetable Center.
Palanisami, K., Meinzen-Dick, R., Giordano, M., 2010. Climate change and water supplies: options for sustaining tank irrigation potential in India. Economic and Political Weekly 14, 183–190.
Paterson, B., 2011. Water. Agriculture and Food: Alberta's Opportunity and Challenges. Fourth McGill Conference on Global Food Security. https://www.mcgill.ca/globalfoodsecurity.
Pathak, P., Sahrawat, K.L., Wani, S.P., Sachan, R.C., Sudi, R., 2009. Opportunities for water harvesting and supplemental irrigation for improving rainfed agriculture in semi-arid areas. In: Wani, S.P., Rockström, J., Oweis, T. (Eds.), Rainfed Agriculture: Unlocking the Potential. Comprehensive Assessment of Water Management in Agriculture Series, CABI, Wallingford, England, pp. 197–221.
Penning de Vries, F.W.T., Acquay, H., Molden, D., Scherr, S.J., Valentin, C., Cofie, O., 2003. Integrated Land and Water Management for Food and Environmental Security. Comprehensive Assessment Research Report 1. International Water Management Institute, Forest Trends and the Global Environmental Facility.
Poddar, R., Qureshi, M.E., Shi, T., 2014. A comparison of water policies for sustainable irrigation management: the case of India and Australia. Water resources management 28 (4), 1079–1094.
Postel, S.L., Daily, G.C., Ehrlich, P.R., 1996. Human appropriation of renewable freshwater. Science 271, 785–788.
Pretty, J., Noble, A., Bossio, D., Dixon, J., Hine, R., Penning de Vries, F.T.W., Morison, J., 2006. Resource-conserving agriculture increases yields in developing countries. Environmental Science and Technology 40, 1114–1119.
Price, A.H., Norton, G.J., Salt, D.E., Ebenhoeh, O., Meharg, A.A., Meharg, C., Davies, W.J., 2013. Alternate wetting and drying irrigation for rice in Bangladesh: is it sustainable and has plant breeding something to offer? Food and Energy Security 2, 120–129.
Qadir, M., Sharma, B.R., Bruggeman, A., Choukr-Allah, R., Karajeh, F., 2007. Non-conventional water resources and opportunities for water augmentation to achieve food security in water scarce countries. Agricultural Water Management 87, 2–22.

Qadir, M., Wichelns, D., Raschid-Sally, L., McCornick, P.G., Drechsel, P., Bahri, A., Minhas, P.S., 2010. The challenges of wastewater irrigation in developing countries. Agricultural Water Management 97, 561–568.

Renault, D., Facon, T., Wahaj, R., 2007. Modernizing Irrigation Management – the MASS-COTE Approach. Mapping System and Services for Canal Operation Techniques. FAO Paper 63.

Rockström, J., Karlberg, L., Wani, S., Barron, J., Hatibu, N., Oweis, T., Bruggeman, A., Farahani, J., Qiang, Z., 2010. Managing water in rainfed agriculture—The need for a paradigm shift. Agricultural Water Management 97, 543–550.

Sayer, J., Sunderland, T., Ghazoul, J., Pfund, J.L., Sheil, D., Meijaard, E., et al., 2013. Ten principles for a landscape approach to reconciling agriculture, conservation, and other competing land uses. Proceedings of the National Academy of Sciences of the United States of America 110 (21), 8349–8356.

Schneider, A.D., Buchleiter, G., Kincaid, D.C., 2000. LEPA irrigation developments. National irrigation symposium. In: Proceedings of the 4th Decennial Symposium, Phoenix, Arizona, USA, November 14–16, 2000, pp. 89–96.

Schottler, S.P., Ulrich, J., Belmont, P., Moore, R., Lauer, J.W., Engstrom, D.R., Almendinger, J.E., 2014. Twentieth century agricultural drainage creates more erosive rivers. Hydrological Processes 28, 1951–1961.

Schultz, B., 2011. Role of tidal lowlands for food production in the humid tropics. In: Madramootoo, C.A. (Ed.), Water Management for Global Food Security. McGill Institute for Global Food Security, McGill University, Montreal, Quebec. 122 pp.

Schultz, B., Zimmer, D., Vlotman, W.F., 2007. Drainage under increasing and changing requirements. Irrigation and Drainage 56, S3–S22.

Shah, T., 2009. Taming the Anarchy: Groundwater Governance in South Asia. Resources for the Future, Washington, DC, USA. International Water Management Institute (IWMI), Colombo, Sri Lanka, pp. 310.

Sharma, B., Amarasinghe, U., Xueliang, C., de Condappa, D., Shah, T., Mukherji, A., Bharati, L., Ambili, G., Qureshi, A., Pant, D., Xenarios, S., Singh, R., Smakhtin, V., 2010. The Indus and the Ganges: river basins under extreme pressure. Water International 35, 493–521.

Siebert, S., Döll, P., Hoogeveen, J., Faures, J.-M., Frenken, K., Feick, S., 2005. Development and validation of the global map of irrigation areas. Hydrology and Earth System Sciences 9, 535–547.

Singh, A.K., Rahman, A., Sharma, S.P., Upadhyaya, A., Sikka, A.K., 2009. Small holders' irrigation-problems and options. Water Resources Management 23, 289–302.

Suhardiman, D., Giordano, M., 2014. Is there an alternative for irrigation reform? World Development 57, 91–100.

Vaez Tehrani, M., Monem, M.J., Bagheri, A., 2013. A system dynamics approach to model rehabilitation of irrigation networks case study: Qazvin irrigation network, Iran. Irrigation and Drainage 62, 193–207.

Venot, J.P., de Fraiture, C., Nti Acheampong, E., 2012. Revisiting Dominant Notions: A Review of Costs, Performance and Institutions of Small Reservoirs in Sub-Saharan Africa. International Water Management Institute, Colombo, Sri Lanka.

Wada, Y., Beek, L., Bierkens, M., 2012. Nonsustainable groundwater sustaining irrigation: a global assessment. Water Resources Research 48, 6.

Wani, S., Sreedevi, T.K., Sudi, R., Reddy, V., Dixin, Y., Li, Z., 2013. Improved Livelihoods and Water Productivity through Integrated Watershed Management – A Case Study from China. Resilient Dryland Systems Report No. 61. International Crops Research Institute for the Semi-Arid Tropics (ICRISAT), Patancheru 502 324, Andhra Pradesh, India. 48 pp.

Wesström, I., Joel, A., Messing, I., 2014. Controlled drainage and subirrigation–A water management option to reduce non-point source pollution from agricultural land. Agriculture, Ecosystems & Environment 198, 74–82.

Wisser, D., Frolking, S., Douglas, E.M., Fekete, B.M., Schumann, A.H., Vörösmarty, C.J., 2010. The significance of local water resources captured in small reservoirs for crop production – a global-scale analysis. Journal of Hydrology 384, 264–275.

Woltering, L., Ibrahim, A., Pasternak, D., Ndjeunga, J., 2011. The economics of low pressure drip irrigation and hand watering for vegetable production in the Sahel. Agricultural Water Management 99, 67–73.

WWAP (World Water Assessment Programme), 2012. The United Nations World Water Development Report 4: Managing Water under Uncertainty and Risk. UNESCO, Paris.

World Bank, 2013. Turn Down the Heat: Climate Extremes, Regional Impacts, and the Case for Resilience. A report for the World Bank by the Potsdam Institute for Climate Impact Research and Climate Analytics. World Bank, Washington, DC.

World Bank, 2007. Emerging public-private partnerships. Water Sector Board Discussion Paper Series Paper No 10. May 2007. In: Irrigation Development and Management, p. 43100.

World Bank, 2006. Directions of Development. Reengaging in Agricultural Water Management: Challenges and Options. http://siteresources.worldbank.org/INTARD/Resources/DID_AWM.pdf.

World Health Organization, 2003. Climate Change and Human Health. Risk and Responses. World Health Organization.

Xie, H., You, L., Wielgosz, B., Ringler, C., 2013. Estimating the potential for expanding smallholder irrigation in sub-Saharan Africa. Agricultural Water Management 131, 183–193.

Zhao, L., Wu, L., Li, Y., Animesh, S., Zhu, D., Uphoff, D., 2010. Comparisons of yield, water use efficiency, and soil microbial biomass as affected by the system of Rice intensification. Communications in Soil Science and Plant Analysis 41 (1), 1–12.

Zimmer, D., Bermond, M., Melliand, M.L., Tato-Serrano, S., 2005. Internal Paper on the Status of Drainage up to 2005. CEMAGREF, Water and Environmental Engineering Department, Antony, France.

Zougmore, R., Mando, A., Stroosnijder, L., Ouédraogo, E., 2004. Economic benefits of combining soil and water conservation measures with nutrient management in semi-arid Burkina Faso. Nutrient Cycling in Agroecosystems 70, 261–269.

Reducing Postharvest Losses

7

Marie-Josée Dumont, Valérie Orsat, Vijaya Raghavan
Bioresource Engineering Department, McGill University, Ste-Anne-de-Bellevue, QC, Canada

1. Introduction

Food is lost or wasted throughout the supply chain, from its initial agricultural production down to its final consumption at the household level. In higher-income (developed) countries, food is largely wasted at the level of the consumer, meaning that it is discarded even if it is likely still suitable for human consumption (FAO, 2011). Additional losses also occur early in the food supply chain in the industrialized regions with inadequate harvest times, unavailability/scarcity of manpower, equipment downfalls, and mismanagement. In lower-income (developing) countries, food is lost principally during the early stages of the food supply chain, due to inadequate handling and distribution channels, while with the high price of food, much less is wasted at the consumer level.

2. Handling of Fresh Fruits and Vegetables

2.1 Harvesting Considerations

At harvest, the quality of crops largely affects their postharvest life. Injuries can easily occur during the mechanical harvesting of fruits and vegetables. These injuries are a serious issue, as these predispose fruits and vegetables to rapid decay caused by increased respiratory rate, increased ethylene production rate, and water loss, and to become a site to host the growth of decay microorganisms.

Delicate crops, such as berries, are manually harvested to prevent mechanical injuries. Harvesters/pickers should be trained to recognize crop maturity stages and indexes such as the skin color, shape, size, aroma, firmness, and leaf changes, among many other traits (Thompson, 2003). Pickers should also be trained to properly snap, cut, or pull fruits and vegetables to minimize the incidence of product injuries. Also, pickers must be careful not to bruise crops once deposited in the basket by throwing the produce in the collecting field containers. Crops should be protected from the sun, as they could suffer from sun overexposure. Harvesting in the early morning or the evening is a good option to protect crops from sun damage or wilting, and it also decreases the energy required to subsequently cool down the crops, as precooling is an essential step in the handling and supply chain of fruits and vegetables prior to reaching the markets. Here, the term "precooling" refers to the removal of field heat prior to further handling. This step slows down the product's metabolism and respiration and therefore reduces deterioration prior to storage, transport, and distribution (Chattopadyay, 2007; Brosnan and Sun, 2001).

Emerging Technologies for Promoting Food Security. http://dx.doi.org/10.1016/B978-1-78242-335-5.00007-X

Figure 1 Liquid-ice precooled broccoli.
Photo courtesy of Prof. Vijaya Raghavan.

2.2 Precooling

Precooling is essential for maintaining an optimal postharvest life and to minimize the impact of the product's senescence on its market quality (Chattopadyay, 2007). It is also considered one of the most economical methods to help preserve the quality and prolong the shelf life of commercial crops (Dehghannya et al., 2010). Precooling can be achieved by several techniques such as room cooling, hydrocooling, vacuum cooling, forced air cooling, and contact icing (Figure 1), among others (Fricke, 2006; Dehghannya et al., 2010; Vigneault et al., 1995).

Some of the characteristics of the most employed precooling methods are described in Table 1. The choice of the precooling method is influenced by the product type. For fruits and vegetables, the precooling method will be influenced by factors, such as the airflow rate, air temperature, relative humidity, the geometry and thermal properties of the produce, and the packing configuration and stacking arrangement (Dehghannya et al., 2010).

As some of the precooling methods presented in Table 1 are energy intensive and thus expensive, great care in handling must be taken, as bruises and mechanical damage occurring after this stage will counteract the benefits of the precooling step.

2.3 Postharvest Treatments

Postharvest treatments generally consist in cleaning, selecting, grading, and packing of the fresh produce. However, some crops necessitate special handling or treatments. Root crops, such as potatoes and some tropical root crops, may require having their roots cured prior to long-term storage. For potatoes, the curing treatment consists of keeping them for 5 days or more at 15–20 °C and 90–95% relative humidity (RH). Curing is a very effective, nonchemical method for reducing decay rot, as it permits the strengthening of damaged areas of the tuber and allows the consolidation of protection against water loss and infection by decay organisms (FAO, 1989). For dry bulb onions, the curing step consists of a drying-out process, which takes place right after harvest, allowing the outer layers to dry into a protective layer for prolonged storage.

Table 1 **Precooling Methods and Description (Barbosa-Cánovas et al., 2005; Kienholz and Edeogu, 2002)**

Method	Simplified Description	Examples
Room cooling	This involves placing crops in a cold room where cold air passing through a fan serves as the refrigeration unit. This method has low or no cost involvement.	Beets, cabbages, peppers, pumpkins, squash, tomatoes, zucchini
Air precooling	Common practice where cold air is used to reduce fruit temperature. Can be performed in a storage room, refrigerator car, tunnel, or forced air cooler.	Mainly fruits
Forced air cooling	Cold air is directed directly through the crop at high velocity. The main disadvantage is the risk of desiccation of the crop.	Beans, brussel sprouts, cabbages, carrots, cauliflower, celery, cucumbers, kohlrabi, leafy greens, lettuce, peas, peppers, potatoes, raspberries, spinach, rhubarb, strawberries
Icing	Ice is used to cool down product packed in boxes. Crushed ice can also be placed directly on top of the crop.	Asparagus, broccoli, leaks, green onions, parsnips, radishes, rutabagas, sweet corn
Hydrocooling	This method consists in submerging crops in cold water. A heat exchanger is necessary to maintain the water temperature. This method permits faster cooling, it helps clean the produce, and if the water is chlorinated, the method can prevent spoilage of the crop. This method is used mainly for precooling vegetables.	Asparagus, beans, broccoli, carrots, cucumbers, kohlrabi, parsnips, radishes, rhubarb, rutabagas, sweet corn
Vacuum cooling	Vacuum cooling is used for crops that do not have a thick wax cuticle. This method cools down crops rapidly due to the evacuation of the air, which causes a rapid evaporation of water located at the surface of the crops. A disadvantage of this method is the reduction in crop weight of 1% for every 5 or 6 °C reduction in temperature.	Lettuce, leeks, Chinese cabbage

Certain crops, such as high-value tropical fruits, must be handled carefully through the cold chain, as they are sensitive to chilling injuries caused by cold storage. Treatment options exist to improve the tolerance of certain crops to chilling injury. Strategies involving physical, chemical, or biotechnological treatments are currently being developed with mixed results (Orsat and Raghavan, 2007). However, the use of heat treatment prior to cold storage is considered as environmentally friendly and is highly effective at enhancing chilling tolerance for a variety of products, such as tomatoes (Aghdam and Bodbodak, 2014; Lu et al., 2010).

There are numerous postharvest treatments that can be applied to fruits and vegetables, and many more will probably become available in future years, as this is a research-intensive area. For example, research is being conducted on the application of chemical agents (e.g., 1-methylcyclopropene) to inhibit the action of ethylene on iceberg lettuce (Tian et al., 2014). In addition, essential oils have been utilized to inhibit the growth of pathogens on baby ginger (Liu et al., 2014) and to control the incidence of decay, prolong storage life, and maintain the internal quality of mango fruits (Kupaei and Garmakhany, 2014). The removal of pesticide residues on fresh lettuce using a chlorine dioxide treatment has also been reported (Chen et al., 2014). Ozonation is another postharvest treatment of growing commercial interest. The use of ozone permits the destruction of pesticides and other chemical residues, and the elimination of microorganisms. Moreover, ozone decomposes quickly in oxygen and therefore does not leave residues on the treated fruits and vegetables (Horvitz and Cantalejo, 2014).

3. Storage of Fresh Fruits and Vegetables

3.1 Refrigerated Storage

Storage of fruits and vegetables is essential as it ensures continuity of supplies, past the market glut, especially in countries located in tropical and subtropical areas where cultivation and harvest periods are much longer than in temperate areas. In temperate regions, with much shorter harvest seasons, storage is even more critical to ensure availability of fresh produce far past the harvest season. At optimal storage conditions, the shelf life of fruits and vegetables ranges from a few days (e.g., blackberries, raspberries, jaboticaba, and lima beans), to weeks (e.g., asparagus, bananas, blueberries, mangos, and spinach) and up to months (e.g., apples, beets, cabbage, and sweet potatoes) (Chattopadyay, 2007). By optimal storage conditions, it is meant that fruits and vegetables will be stored at their product-specific optimum temperature and optimum relative humidity. When stored in mixed storage, it is important, at a minimum, to maintain product grouping in order to respect their requirements as per their compatibility group. Seven compatibility groups for storage conditions have been categorized as follows (Kitinoja and Kader, 2002):

- Group 1: Storage temperature 0–2 °C, relative humidity 90–95%. Particularity: Many products produce ethylene. E.g., apricots, leeks, pomegranates, turnips.
- Group 2: Storage temperature 0–2 °C, relative humidity 95–100%. Particularity: Many products are sensitive to ethylene. E.g., asparagus, bok choy, kiwifruits, water chestnuts.
- Group 3: Storage temperature 0–2 °C, relative humidity 65–75%. Particularity: Moisture will damage these products. E.g., garlic, dry onions.

Figure 2 Perforated piping system to ensure aeration of a potato storage pile to prevent moisture buildup and spoilage (seen on the right).
Photos courtesy of Prof. Vijaya Raghavan.

- Group 4: Storage temperature 4.5 °C, relative humidity 90–95%. E.g., cactus leaves, clementines, cranberries, oranges (California and Arizona), and yucca roots.
- Group 5: Storage temperature 10 °C, relative humidity 85–90%. Particularity: Many products are sensitive to ethylene and to chilling injury. E.g., beans, chayote, eggplants, peppers, and olives.
- Group 6: Storage temperature 13–15 °C, relative humidity 85–90%. Particularity: Many products produce ethylene and are sensitive to chilling injury. E.g., avocados, bananas, coconuts, ginger roots, grapefruits, pineapple, pumpkins, and soursops.
- Group 7: Storage temperature 18–21 °C, relative humidity: 85–90%. E.g., jicama, pears (for ripening), tomatoes (mature green), watermelon, white sapote, and yams.

Temperature management and time are the most important parameters associated with product freshness (Aung and Chang, 2014). To maintain the quality of fresh fruits and vegetables and to prolong their shelf life, low temperature storage is the main strategy used. The efficiency of this method mainly comes from the decrease in the respiration rate of the fresh produce, which decreases the production and sensitivity to ethylene (Valero, 2010) and decreases fungal disease incidence (Aghdam and Bodbodak, 2014). Certainly for foods that must be stored, careful attention must be taken not to deviate to a warmer temperature, as microbial growth may likely occur and cause spoilage of foods, and in certain cases may cause food poisoning (Carullo et al., 2009). The relative humidity in the environment of the commodity should be carefully monitored, along with the temperature, in order to prevent moisture buildup or condensation on the produce surface, which would favor decay (Figure 2).

3.2 Maintaining the Cold Chain

Fruits and vegetables need to be stored at low temperatures as soon as possible following harvesting to reduce their metabolic activity. Therefore, precooling techniques, as described earlier in Table 1, such as forced air cooling, hydrocooling, contact icing, vacuum cooling, and cryogenic cooling, can be used to rapidly decrease the heat associated with the field temperature as well as decrease the respiration rate of fresh produce (Valero, 2010). Afterward, the precooled crops should be stored and distributed within a refrigerated structure. It is important to note that low-temperature treatments can cause chilling injuries, which consist in the formation of detrimental ice crystals inside the tissues that may cause cellular breakdown. Tropical and subtropical fruits and vegetables are more susceptible to develop low-temperature-related injuries, and these can occur at temperatures well above those at which the crop would freeze (Aghdam and Bodbodak, 2014; Thompson, 2003). These injuries lead to a variety of quality characteristics such as an uneven or abnormal ripening, an increase in water loss, discoloration and surface pitting, internal browning, tissue breakdown, etc. (Serrano et al., 1996). However, the susceptibility to such chilling injuries can be lessened by subjecting the produce to a short heat treatment prior to cold storage (Lu et al., 2010). The chilling tolerance is affected by a produce reaction triggered by several mechanisms such as an increase in the unsaturated to saturated fatty acid ratio, which increases the integrity of the membrane; an increase in enhancement of the expression and accumulation of heat shock protein expression; an increase in the antioxidant system activity and sugar metabolism; and finally, an enhancement in the arginine pathway, which leads to the accumulation of signaling molecules, which improves the chilling tolerance. Heat treatments can also alter the phenylalanine ammonia-lyase and polyphenol oxidase enzymatic activities (Aghdam and Bodbodak, 2014).

3.3 Controlled or Modified Atmosphere Storage

Controlled or modified atmosphere storage can be used as a complement to proper temperature and relative humidity management (Kitinoja and Kader, 2002). Controlled atmosphere storage refers to a continuously controlled gas atmosphere, while modified atmosphere storage refers to a gas composition that is initially modified. The gas composition within a modified atmosphere storage will change over time due to the respiration rate of food products and the permeability of the packaging (if any) surrounding the food products (Jayas and Jeyamkondan, 2002). The primary benefit of lowering O_2 and increasing CO_2 surrounding the produce is the effect it has on lowering the respiration/metabolic rate of the commodity, leading to slowing down of the natural senescence process. Other effects of modified O_2 and CO_2 levels are summarized in Table 2 (Thompson, 1998). From Table 2, careful attention must be taken to maintain a good control on the precise levels of modified gases, since too high levels

Table 2 Some effects of modified O_2 and CO_2 levels on crops (Thompson, 1998)

Modified O_2 Levels	Modified CO_2 Levels
Reduced respiration rate	Decreased discoloration levels
Reduced substrate oxidation	Decreased synthetic reaction in climacteric fruits
Delayed ripening of climacteric fruits	Inhibition of some enzymatic reactions
Prolonged storage life	Decreased production of some organic volatiles
Reduced rate of production of ethylene	Modified metabolism of some organic acids
Reduced degradation rate of soluble pectin	Inhibition of the effects of ethylene
Formation of undesirable odors and flavors	Retarded fungal growth
Altered texture	Production of off-flavors
Development of physiological disorders	Development of physiological disorders

of CO_2 or O_2 will likely injure fruits and vegetables. Examples of optimized modified atmosphere conditions are described in Table 3 (Kader, 1994).

Another great benefit of controlled atmosphere storage is related to the reduction of food deterioration from pests and diseases, since higher CO_2 levels generally have a negative impact on the development and growth of microorganisms. Therefore, storing fruits and vegetables under a controlled atmosphere may significantly reduce the amount of postharvest chemicals used for their protection against insect and microorganisms (Thompson, 1998).

3.4 Alternative Storage Techniques

Cold storage and controlled atmosphere storage are effective but expensive storage methods. An alternative suggestion is the choice of using natural storage options as the most convenient and cheapest way of storing fruits and vegetables. Natural storage is defined as the simple practice of leaving crops in the soil until they are ready for the market. This can be applied to roots (e.g., carrots and cassava) and tubers (e.g., potatoes) (Chattopadyay, 2007). Another method useful for hearty fruits and vegetables is storage in earthen structures and underground caves, where lower temperatures and higher relative humidity can be maintained.

Alternative methods such as evaporative cooling may be employed for harvested crops to help extend their shelf life at a low cost. Countries such as India and other emerging countries have benefitted from this technology, as it is low-cost, low-energy, and environmentally friendly as compared to refrigeration. Low temperature and high relative humidity (>90%) conditions are obtained due to an adiabatic exchange of heat as air passes through a water-saturated surface, picking up moisture, which reduces the temperature. The cooling chambers have the advantages of not requiring mechanical or electrical energy input, can be built from locally available materials, do not require special skills to be operated, and fruits and vegetables can be successfully stored for 3–5 days without significant loss. However, the main disadvantage of this system would be the need for a constant supply of water (Basediya et al., 2013).

4. Handling and Storage of Grains and Cereals

Cereals are the primary source of food for most of the world's population. Maize, rice, and wheat together account for over 80% of the world cereal production. The latest assessment from the Food and Agriculture Organization of the United Nations (FAO) puts world cereal production in 2014 at nearly 2480 million tons (FAO, 2014b). Other important cereal productions include barley, rye, oats, millet, and sorghum. The increase in human food consumption of cereals is seen to keep pace with the population growth.

4.1 Harvesting Considerations

Time of harvest will govern the quality of the produce and will dictate the subsequent handling that will be required. There appears to be a persistent lack of knowledge

Table 3 Examples of Modified Atmosphere Conditions and Benefits for Selected Tropical Fruits (Kader, 1994)

	Avocados		Papayas		Bananas	
	Reduced O_2	Increased CO_2	Reduced O_2	Increased CO_2	Reduced O_2	Increased CO_2
Beneficial levels	2–5%	3–10%	2–5%	5–8%	2–5%	2–5%
Benefits	Delay in ripening, reduced rate of CO_2 and C_2H_4 production	Delay in softening, reduced chilling injury symptoms	Delay in ripening	Firmness retention	Delay in ripening	Delay in ripening
Potential for benefits	Good	Good	Slight to moderate	Slight to moderate	Very good	Very good
Injury symptoms	Off-flavor, internal flesh browning	Skin browning, off-flavor	Failure to ripen, off-flavors	Off-flavors, may aggravate chilling injury at <12 °C	Discoloration, failure to ripen, off-flavors	Green fruit softening, undesirable texture and flavor
Potential for injury	Moderate	Moderate	Moderate	Moderate	High	Moderate to high
Optimum storage temperature	10 °C		12 °C		14 °C	

pertaining to the established maturity index of certain grains and cereals, and the required planning of harvesting to meet the maturity index for local or export markets and long-term storage. The quality of the grain and its moisture content at harvest will dictate how it should be handled.

4.2 Handling and Drying Requirements

In temperate climates, about 126 kg of water must be removed per MT of coarse grain harvested for use in the long term, and the corresponding energy requirements range from 0.38 to over 0.63 GJ per MT of grain, depending on the type of commercial dryer used, the required moisture reduction, and the operating conditions (Raghavan and Sosle, 2006).

In first handling of the grain, aeration may be the adequate tool to implement. Aeration consists of moving air through the grain in order to equalize both the grain temperature and moisture and prevent moisture build-up. Aeration can be considered to cool the grain after drying, to remove odors, or to distribute fumigants through the grain mass. This operation is usually carried out in a storage bin that is equipped with a fan, duct system, and perforated floor along with exhaust vents to provide escape of the moisture-filled air (Figure 3).

Drying can also be done in the storage bin by ventilating with air that is heated a minimum of 5–10 °C above ambient temperature. In-bin drying is adequate for limited grain quantities to remain effective (Brooker et al., 1974). This method usually requires continuous operation of the ventilation system over a period of one to three weeks. Drying can be conducted using an independent drying unit prior to storing the grain. This is often the choice offering the greatest flexibility in grain handling. Drying recommendations for selected grains and cereals are presented in Table 4.

Where the weather permits, natural and sun drying are valid options. To be most effective, sun or natural drying should be done in shallow layers. This form of drying may be achieved by spreading the produce in a layer on the ground or on wire-bottom

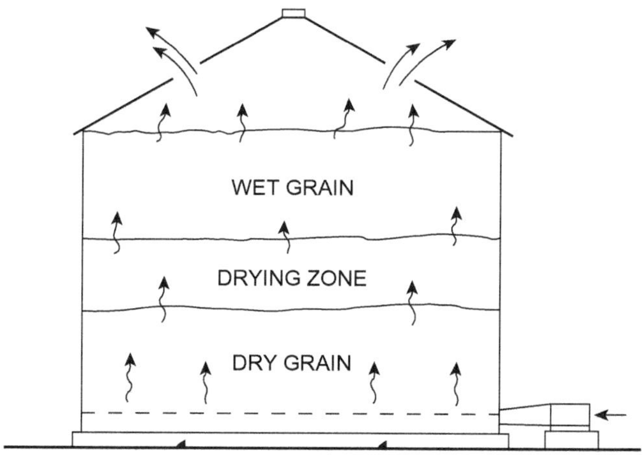

Figure 3 Unheated air drying (Alberta Agriculture, 2015).

Table 4 **Recommendations for Drying Grain with Natural Air and Heated Air (Raghavan and Sosle, 2006)**

	Ear Corn	Shelled Corn	Wheat	Oats	Barley	Sorghum	Soy	Rice
Max. moisture content of crop at harvesting for satisfactory drying								
With natural air (%)[a]	30	25	20	20	20	20	20	25
With heated air (%)[a]	35	35	25	25	25	25	25	25
Max. moisture content[b] of crop for safe storage in a tight structure (%)[a]	13	13	13 (12)[c]	13	13 (12)[c]	12	11	12
Max. relative humidity for natural air drying to safe storage level (%)	60	60	60	60	60	60	65	60
Max. temp. of heated air when crop is								
Used for seed (°C)	43	43	43	43	41	43	43	43
Sold for commercial use (°C)[a]	54	54	60	60	41	60	49	43
Animal feed (°C)[a]	82	82	82	82	82	82		

[a]Moisture contents on wet basis: (a) higher temperatures than those listed may be used when the corn is dried under carefully controlled conditions such that temperature of the kernels does not exceed 54 °C at any time; (b) if there is any possibility that the crop may be sold, use the lower temperature as listed for commercial use.

[b]If the products are to be stored for long periods, the moisture content should be 1–2% lower than shown in this tabulation.

[c]Seed.

trays that are supported above the ground. Cribs (storage structure allowing natural ventilation) may also be constructed for drying maize on the cob or unthreshed legumes and cereals. The crib structures are usually oriented so that their long axis faces the prevailing wind.

4.3 Alternate Storage Systems

The absence of adequate storage structures for grain storage and the nonexistence of product management strategies usually force the small grain producers to market their produce immediately following harvest. This often leads to a market glut, which brings down prices. Without stored produce, the producers have no bargaining power and no reserve to feed their family (Tefera et al., 2011). Small-scale storage options are available with metal containers or metal silos for household storage (Figure 4); while for medium-size capacity, mud or brick storage is an option for village-level storage needs (Figure 5); and of course for large-capacity storage, the larger grain

Figure 4 Small-scale storage option typically available in India.
Photo courtesy of Prof. Vijaya Raghavan.

Figure 5 Village-level brick grain storage (under construction, at left).
Photos courtesy of Prof. Vijaya Raghavan.

silos, commonly found at grain distribution centers, are the option. In most countries where the metal silo has been introduced, the silo has created a positive impact among stakeholders directly or indirectly associated with favoring grain production and subsequent availability of the grain for consumption (Tefera et al., 2011).

5. Quantitative and Qualitative Losses

Worldwide, 1.3 billion tons per year of food produced will never be consumed by humans (Liu et al., 2013). Fruits and vegetables account for approximately one-third of these losses, and grains account for about one-fifth (Kader, 2005; de Lucia and Assennato, 1994; Liu et al., 2013).

Food loss may be defined as a decrease of edible food mass that occurs through the different steps of the supply chain (e.g., production, postharvest, processing) and also includes the waste component, which represents edible goods discarded at the end of the food chain (retail and final consumption) (Buzby et al., 2014; Gustavsson et al., 2011; Liu et al., 2013). These losses are quantitative or qualitative. The latter refer to the loss in edibility, nutritional quality, caloric value, and consumer acceptability (Kader, 2005). Fresh-cut fruits and vegetables are examples of foods that are highly susceptible to qualitative losses. The abiotic stress caused by the injury leads to discoloration reactions at the freshly cut surface, an increase of respiration rate and ethylene evolution, a loss of flavor and texture, the development of off-odors, and tissue softening, among other symptoms (Hodges and Tolvonen, 2008). Table 5 summarizes the major causes of food losses at the different steps of the food supply chain (Liu et al., 2013).

The average distributions of percentage losses between developed and developing countries are the same, reaching about 32% (Prusky, 2011). To illustrate this, it was estimated that Europe, North America, Oceania, and industrialized Asia have a food loss rate of 35%, while North Africa, West and Central Asia, and Latin America have an estimated food loss rate of 25–30%. China seems to perform exceedingly well, having an average food loss rate of 19% (Liu et al., 2013). The main difference between developed and developing countries is the stages at which the losses occur. The average losses occurring from production to retail sites are estimated at 12% in developed countries, while they reach around 22% in developing countries. On the other hand, the average loss or waste of still perfectly edible foodstuff at the retailer and consumer level reach on average 20% in developed countries, while in general, it only reaches about 10% in developing countries. This is due to rigid quality controls and "best before" standards in developed countries or due to the consumer who wastes food, as it was not consumed prior to the expiration date or discarded after a meal (Prusky, 2011; Hodges et al., 2011). Overall, it was shown that Europeans and North Americans generate about 280–300 kg/person/annum of food wastes as compared to 170 kg/person/annum in developing countries, such as those located in Sub-Saharan Africa (Nahman and de Lange, 2013). Table 6 presents the losses of grains, fruits, and vegetables in the United States at the retail compared with the consumer levels (Buzby et al., 2014).

Table 5 Causes of Food Losses and Wastes at Different Steps of the Food Supply Chain (Liu et al., 2013)

	Harvest	Storage	Transportation	Processing	Consumer
Causes of loss	1. Pests 2. Mechanical damage 3. Spillage 4. Crop sorting	1. Spillage 2. Degradation 3. Pests 4. Drying out	1. Spillage 2. Degradation 3. Spoiling 4. Bruising	1. Spillage 2. Degradation 3. Crops Spoiling	1. Inedible food discarded 2. Edible food discarded with inedible 3. Discarded in packaging 4. Throw away after expiration dates

Table 6 Food Losses and Associated Dollar Value at the Retail and Consumer Levels in the United States (Buzby et al., 2014)

	Retail			Consumer		
	Billion kg	**%**	**Billion $**	**Billion kg**	**%**	**Billion $**
Grain products	15.8	12	4.3	24.9	19	6.9
Fruits	19.8	9	5.8	27.5	19	14.1
Vegetables	15.4	7	9.6	40.0	22	20.4

Table 7 Food Loss (% mass) Occurring at Different Stages of the Food Supply Chain in Sub-Saharan Africa (Nahman et al., 2012; Gustavsson et al., 2011)

	Cereals	Roots and Tubers	Fruits and Vegetables
Agricultural production	6%	14%	10%
Postharvest handling and storage	8%	18%	9%
Processing and packaging	3.5%	15%	25%
Distribution	2%	5%	17%
Consumption	1%	2%	5%
Total	20.5%	54%	66%

In less-developed countries, the main cause of food products losses comes from limitations in the harvesting techniques and unavailability of appropriate storage and cooling facilities, resulting in biological spoilage early on in the distribution chain. Additional limitations can be found in the distribution infrastructure, product packaging, and the food marketing systems (Hodges et al., 2011; Nahman and de Lange, 2013; Parfitt et al., 2010; Nahman et al., 2012). To illustrate this, Table 7 presents losses of cereals, roots and tubers, and fruits and vegetables in Sub-Saharan Africa at different stages of the food supply chain (Nahman et al., 2012; Gustavsson et al., 2011).

From Tables 6 and 7, it can be seen that products considered less perishable or nonperishable, owing to their lower moisture content (e.g., tubers, roots, grains, and cereals), can sustain significant losses as well. In less developed countries, grain loss is principally associated with inadequate postharvest handling, decay, and storage pests. The average grain production and associated losses for 16 countries in east and southern Africa are presented in Table 8. Farmers in developing countries often use more traditional postharvest handling and processing methods and practices, which at times can lead to greater product losses. For example, sun drying may generate losses from insects and molds, especially when drying is insufficient due to the vagaries of weather. However, the implementation of a more mechanized

Table 8 Average Grain Production and Losses for 16 Countries in East and Southern Africa for the Year 2007 (Hodges et al., 2011)

Grain Type	Annual Production (million tonnes)	Average % Weight Loss Estimate	Value of Losses in U.S. $ million
Maize	27.01	17.5	920
Sorghum	4.72	11.8	139
Millet	1.67	11.7	60
Wheat	5.25	13.0	187
Barley	1.71	9.9	48

Table 9 Estimated Qualitative and Quantitative Losses from the Postharvest Chain for Rice in South Asia (Hodges et al., 2011)

Weight Loss in Traditional Postharvest Chain		Weight Loss in Mechanized Postharvest Chain	
Cutting and handling	1–5%	Combined harvesting	1–5%
Manual threshing	1–5%	Machine threshing	1–5%
Sun drying	3–5%	Mechanical drying	1–2%
Open storage	5–10%	Scaled storage	1–2%
Village milling	20–30%	Commercial milling	5–30%
Minimum waste attainable	30%	Minimum waste attainable	9%
Maximum waste attainable	55%	Maximum waste attainable	44%

postharvest chain could help decrease grain losses (Table 9) (Hodges et al., 2011; de Lucia and Assennato, 1994).

The fact that a quarter of all produced food is lost represents a major concern, as the global population is estimated to reach 9.1 billion by 2050 (Hodges et al., 2011; Kummu et al., 2012). In addition to being losses of food, these losses create a significant waste of other resources. For example, 24% of agricultural freshwater resources, 23% of cropland area, and 23% of fertilizer are used in vain in the production of this food that ends up as food loss (Kummu et al., 2012). It has been hypothesized that by lowering the percentage of losses, using existing postharvest practices, more food would remain available for consumption by an extra billion people (Kummu et al., 2012).

6. Packaging and Distribution Chain

The FAO estimated that globally, 1.3 billion tons per year of the food produced for human consumption get lost or wasted in some way (Gustavsson et al., 2011). This represents one-third of the global food production. Interestingly, the FAO also reported

Table 10 Average Import of Packaging Materials from 2005 to 2009 (U.S. $) (FAO, 2014a)

Packaging Material	Underdeveloped Economy	Developing Economy	Developed Economy	Total
Glass	6	956	590	1556
Paper	7	1892	2048	3959
Plastic	24	4670	4882	9589
Wood	0	155	123	285

that the percentage of these losses is the exact same for both developing and industrialized countries. The difference in these losses is that they do not occur at the same stage in the food distribution chain. As such, about 40% of food losses occur principally at the postharvest and processing levels in developing countries. In contrast, developed countries experience 40% of food losses, principally at both the retail and consumer levels (Gustavsson et al., 2011). Adequate packaging can help reduce these losses, especially at the retail and consumer levels.

The selection of appropriate packaging can significantly reduce food losses and can increase food safety throughout the food chain stages by protecting the goods from chemical, biological, and physical aggressions (Marsh and Bugusu, 2007). The food and beverage sector accounts for the majority of the global packaging industry. Food alone accounts for 50% of the global packaging industry, representing a value of U.S. $380 billion, while the beverage sector accounts for 19% of the industry (FAO, 2014a; Marsh and Bugusu, 2007). The market value of the packaging industry in underdeveloped, developing, and developed economies is presented in Table 10 (FAO, 2014a).

Chemical alteration of foods occurs through compositional changes caused by exposure to gases (e.g., oxygen) and light, but also due to any alteration in the moisture content of the food. Microorganisms, insects, rodents, and the like are common biological aggressors from which foods require protection by means of packaging. In addition, a barrier to biological attacks can also be used to maintain adequate conditions for senescence control. Finally, packaging can also offer a physical protection that guards the goods from shocks and vibrations during transportation, distribution, and storage.

The most utilized packaging materials are glass, paper, plastic, metal, and wood (Raheem, 2013). Glass offers an excellent barrier against chemical aggressors such as gas and vapors, can protect against light if colored, and is resistant to the food's pH. Its reusability and recyclability make glass an environmentally friendly material. Because of its rigidity, glass may be perceived as a protector against shocks; however, its brittleness renders this packaging material susceptible to breakage. Most probably the main disadvantage of this material is its heavy weight, which adds to transportation costs (Marsh and Bugusu, 2007).

Paper and paperboard are mostly used as a physical protector for goods, as they are not suitable to preserve food for a long period of time. They are used in corrugated boxes, as milk cartons, folding cartons, bags and sacks, and wrapping paper (Berger and Welt, 2005; Raheem, 2013). In order to improve their resistance to chemical aggressors, coatings such as waxes or polymeric materials can be used to enhance their poor barrier properties to oxygen, carbon dioxide, and water vapors (Raheem, 2013). Different examples of chemically modified papers and paperboards, as well as their uses, are found in Table 11 (Marsh and Bugusu, 2007).

Similarly to paper and paperboards, wood containers offer a physical protection for various commodities, especially at the distribution and shipment stage. Wooden pallets are universally used at the distribution stage. Other uses of wood containers include wooden boxes, crates, and baskets used in the harvesting and postharvest handling of fruits and vegetables. Wooden barrels are still utilized for ripening of a number of alcoholic beverages (Lee et al., 2008).

Metals, particularly aluminum, steel, and tin, are used in the fabrication of several types of food containers, such as those for beer, soft drinks, and processed foods in the form of cans, drums and pails, aerosol containers for whipped cream, tubes, open trays, caps, closures, and lids (Oldring and Nehring, 2007). Metals are often coated when in contact with the food in order to prevent discoloration (e.g., from iron), corrosion when in contact with acidic food, or to prevent sulfur stains from fish, meats, and soups. External coatings can also be applied to protect the metal from the environment, to support the application of labels, etc. Metal containers, when sealed, offer an excellent protection against chemical, biological, and physical aggressions for effective long-term storage.

Innovation in packaging is driven by several factors, namely convenience, consumer desire for minimally processed foods, changes in retail and distribution practices, food service needs, trends toward more sustainable packaging, and demands for the global distribution of food (Floros et al., 2010). Undoubtedly, polymeric packaging has been evolving at a very fast pace in the past recent years. Plastic packaging is traditionally used because of its flexibility, light weight, and chemical resistance while offering a tremendous range of physical and optical properties, shapes, and colors. The thermosealability and microwavability have rendered this type of packaging ideal for ready-to-eat meals. Plastic polymers are also used in many other applications, such as yoghurt and beverage containers, packaging films, trays, etc. The most utilized polymeric matrices are based on polyolefins, polyesters, polyvinyl chloride, polyamide, and ethylene vinyl alcohol (Marsh and Bugusu, 2007). The need for laminates and co-extrusion is often required when polymers are used for perishables in order to reach an acceptable level of physical properties such as barrier to gas and moisture, especially in modified atmosphere packaging applications.

As stated, polymeric packaging evolved significantly to satisfy many of the consumers' desires. One of the fastest growing convenience food products in history is fresh-cut fruits and vegetables (Ayala-Zavala et al., 2008). However, fresh-cut fruits and vegetables do respire and transpire, which is highly accountable for a great incidence of decay and physiological changes (Fonseca et al., 2002) and therefore can lead to significant discoloration such as browning, softening, surface dehydration, water loss, translucency, off-flavor and off-odor development, and microbial spoilage (Rojas-Grau et al., 2009). The use of modified atmosphere packaging (MAP) to reduce the oxygen level and increase carbon dioxide concentration, has provided several positive effects, such as a decrease in softening rate, an improved pigment

Table 11 Types of Paper and Paperboards Used in Food Packaging (Marsh and Bugusu, 2007)

Material	Types		Characteristics	Uses
Paper	Kraft paper		Produced by a sulfate treatment process and available in different forms: natural brown, unbleached, bleached, white, and heavy duty.	Bags, wrapping and packaging of flour, sugars, dried fruits and vegetables.
	Sulfite paper		This paper is glazed to improve resistance to oil and wet strength. Lighter and weaker than kraft paper.	Packaging of biscuits and confectionaries.
	Greaseproof paper		Manufactured by the beating process. Resistant to oil but not resistant to wet agents.	Wrapping snack foods, cookies, candies, and other types of oily foods.
	Glassine		Greaseproof, dense, highly smooth, and glossy finish.	Used as liner for biscuits, cooking fats, fast foods, and baked products.
	Parchment paper		Impervious to oil and water. Made by acid-treated pulp. Poor barrier to air and moisture.	Used for fat packaging.
	Laminates		Based on kraft and sulfite pulp. Can be laminated with plastic or aluminum to improve selected properties, such as barrier to gas and water.	Used to package dried products like soups, herbs, and spices.
Paperboard	White board		May be laminated with wax or polyethylene for heat-sealability.	Used as the inner layer of carton. Only type of paperboard that can be in direct contact with food.
	Solid board		Possesses strength and is durable. When laminated with polyethylene, it is used to manufacture liquid cartons (milk boards).	Used to package fruit juices, milk, and soft drinks.
	Chipboard		Made from recycled paper and may contain impurities. Often lined with white board. The cheapest of the category.	Used as outer layer of cartons used for packaging cereals, teas and the like.
	Fiberboard	Solid	Inner white board layer and outer kraft layer.	Used to package dry products such as coffee and milk powder when laminated with plastics or aluminum.
		Corrugated	Made with two layers of kraft paper and a central corrugated material.	Shipment of retail food products and bulk food due to its high resistance to impact abrasion and crushing.

retention, and an overall improvement in the quality preservation of the products (Han, 2005). By reducing the oxygen level surrounding the fruits and vegetables, the respiration rate and several metabolic activities are reduced. This controls a reduction in the ripening rate, and consequently a reduction in the production of ethylene (for certain products)—a by-product of the ripening process. This is beneficial, as ethylene itself accelerates the ripening process, which is one cause of food spoilage (Jayas and Jeyamkondan, 2002). On the other hand, the reduction of the oxygen level below a certain limit can result in anaerobic respiration and the production of ethylene, which will in turn result in a fast deterioration of the product quality, due to an accumulation of acetaldehyde, ethanol, and organic acids. Under these conditions, off-flavors will be developed, and the products will suffer from discoloration and tissue softening. Note that the development of *Clostridium botulinum* is also a potential hazard, as this microorganism grows under anaerobic conditions (Han, 2005).

The use of polymeric packaging is, however, of growing concern among the population, as packaging is considered as environmentally unfriendly and potentially damaging. Therefore, efforts have been concentrated on the development of bio-based packaging, and research is demonstrating the effectiveness or potential of bio-nanocomposites (Rhim et al., 2013) and that certain biomass, such as starches (Averous, 2004) and protein-based packaging (Shi and Dumont, 2014), can be effective. But most importantly, a review demonstrated that the life-cycle assessments indicate that food losses constitute a greater share of the harmful environmental impacts than the production of the package containing the food (Gronman et al., 2013).

7. Conclusions

Global strategies and policies will need to be developed in order to successfully minimize food losses at the field level, during harvest and postharvest stages, and through value addition and crop processing, food processing, handling of crop residues and wastes, and by-product recovery. Consolidation of food security will occur through meaningful and deliberate interactions between the science that we know and the society that we want. Creation of trained human resources that can bring about those interactions, throughout the food chain, is vital to integrate a multidimensional technological approach to food resource management, leading to self-reliance and long-lasting impacts on food security.

References

Aghdam, M.S., Bodbodak, S., 2014. Postharvest heat treatment for mitigation of chilling injury in fruits and vegetables. Food and Bioprocess Technology 7, 37–53.

Alberta Agriculture and Rural Development, 2015. www1.agric.gov.ab.ca/$department/deptdocs.nsf/all/crop1204.

Aung, M.M., Chang, Y.S., 2014. Temperature management for the quality assurance of a perishable food supply chain. Food Control 40, 198–207.

Averous, L., 2004. Biodegradable multiphase systems based on plasticized starch: a review. Journal of Macromolecular Science-Polymer Reviews C44, 231–274.

Ayala-Zavala, J.F., Del-Toro-Sanchez, L., Alvarez-Parrilla, E., Gonzalez-Aguilar, G.A., 2008. High relative humidity in-package of fresh-cut fruits and vegetables: advantage or disadvantage considering microbiological problems and antimicrobial delivering systems? Journal of Food Science 73, R41–R47.

Barbosa-Cánovas, G.V., Fernández-Molina, J.J., Alzamora, S.M., Tapia, M.S., López-Malo, A., Chanes, J.W., 2005. Handling and Preservation of Fruits and Vegetables by Combined Methods for Rural Areas. FAO, Rome.

Basediya, A.L., Samuel, D.V.K., Beera, V., 2013. Evaporative cooling system for storage of fruits and vegetables – a review. Journal of Food Science and Technology-Mysore 50, 429–442.

Berger, K.R., Welt, B., 2005. A Brief History of Packaging. Institute of Food and Agricultural Sciences. University of Florida, USA.

Brooker, D.B., Bakker-Arkema, F.W., Hall, C.W., 1974. Grain drying systems. In: Drying Cereal Grains. AVI Publishing Company, Inc., Westport, Connecticut, pp. 145–184.

Brosnan, T., Sun, D.W., 2001. Precooling techniques and applications for horticultural products – a review. International Journal of Refrigeration-Revue Internationale Du Froid 24, 154–170.

Buzby, J.C., Wells, H.F., Hyman, J., February 2014. The Estimated Amount, Value, and Calories of Postharvest Food Losses at the Retail and Consumer Levels in the United States, EIB-121, U.S. Department of Agriculture, Economic Research Service.

Carullo, A., Corbellini, S., Parvis, M., Vallan, A., 2009. A wireless sensor network for cold-chain monitoring. IEEE Transactions on Instrumentation and Measurement 58, 1405–1411.

Chattopadyay, S.K., 2007. Handling, Transportation and Storage of Fruits and Vegetables. [Online] Gene-Tech Books, New Delhi. Available: http://site.ebrary.com/id/10416779.

Chen, Q.Q., Wang, Y., Chen, F., Zhang, Y.Y., Liao, X.J., 2014. Chlorine dioxide treatment for the removal of pesticide residues on fresh lettuce and in aqueous solution. Food Control 40, 106–112.

Dehghannya, J., Ngadi, M., Vigneault, C., 2010. Mathematical modeling procedures for airflow, heat and mass transfer during forced convection cooling of produce: a review. Food Engineering Reviews 2, 227–243.

de Lucia, M., Assennato, D. (Eds.), 1994. Agricultural Engineering in Development – Post-harvest Operations and Management of Foodgrains. FAO.

FAO, 1989. Prevention of Post-harvest Food Losses Fruits, Vegetables and Root Crops a Training Manual. Food and Agriculture Organization of the United Nations, Rome.

FAO, 2011. Global Food Losses and Food Waste – Extent, Causes and Prevention. Food and Agriculture Organization of the United Nations, Rome.

FAO, 2014a. Appropriate Food Packaging Solutions for Developing Countries (authored by M.M. Nerlita, M.A. Dorado and R. Van Otterdijk). Food and Agriculture Organization of the United Nations, Rome.

FAO, 2014b. FAO Cereal Supply and Demand Brief. Food and Agriculture Organization of the United Nations, Rome.

Floros, J.D., Newsome, R., Fisher, W., Barbosa-Canovas, G.V., Chen, H.D., Dunne, C.P., German, J.B., Hall, R.L., Heldman, D.R., Karwe, M.V., Knabel, S.J., Labuza, T.P., Lund, D.B., Newell-Mcgloughlin, M., Robinson, J.L., Sebranek, J.G., Shewfelt, R.L., Tracy, W.F., Weaver, C.M., Ziegler, G.R., 2010. Feeding the world today and tomorrow: the importance of food science and technology an IFT scientific review. Comprehensive Reviews in Food Science and Food Safety 9, 572–599.

Fonseca, S.C., Oliveira, F.A.R., Brecht, J.K., 2002. Modelling respiration rate of fresh fruits and vegetables for modified atmosphere packages: a review. Journal of Food Engineering 52, 99–119.
Fricke, B.A., 2006. Precooling fruits & vegetables. Ashrae Journal 48, 20–22.
Gronman, K., Soukka, R., Jarvi-Kaariainen, T., Katajajuuri, J.-M., Kuisma, M., Koivupuro, H.-K., Ollila, M., Pitkanen, M., Miettinen, O., Silvenius, F., Thun, R., Wessman, H., Linnanen, L., 2013. Framework for sustainable food packaging design. Packaging Technology and Science 26, 187–200.
Gustavsson, J., Cederberg, C., Sonesson, U., Van Otterdijk, R., Meybeck, A., 2011. Global Food Losses and Food Waste: Extent Causes and Prevention. Food and Agriculture Organization of the United Nations.
Han, J.H., 2005. Innovations in Food Packaging. Elsevier.
Hodges, D.M., Tolvonen, P.M.A., 2008. Quality of fresh-cut fruits and vegetables as affected by exposure to abiotic stress. Postharvest Biology and Technology 48, 155–162.
Hodges, R.J., Buzby, J.C., Bennett, B., 2011. Postharvest losses and waste in developed and less developed countries: opportunities to improve resource use. Journal of Agricultural Science 149, 37–45.
Horvitz, S., Cantalejo, M.J., 2014. Application of ozone for the postharvest treatment of fruits and vegetables. Critical Reviews in Food Science and Nutrition 54, 312–339.
Jayas, D.S., Jeyamkondan, S., 2002. Modified atmosphere storage of grains meats fruits and vegetables. Biosystems Engineering 82, 235–251.
Kader, A.A., 1994. Modified and Controlled Atmosphere Storage of Tropical Fruits.
Kader, A.A., 2005. Increasing food availability by reducing postharvest losses of fresh produce. In: Mencarelli, F., Tonutti, P. (Eds.), Proceedings of the 5th International Postharvest Symposium, vols 1–3.
Kienholz, J., Edeogu, I., 2002. Fresh Fruits & Vegetables Pre-cooling for Market Gardeners in Alberta. Alberta Agriculture, Food and Rural Development, Edmonton, Alberta, Canada.
Kitinoja, L., Kader, A.A., 2002. Small-scale Postharvest Practices a Manual for Horticultural Crops, fourth ed. Postharvest Technology Research and Information Center, California.
Kummu, M., de Moel, H., Porkka, M., Siebert, S., Varis, O., Ward, P.J., 2012. Lost food, wasted resources: global food supply chain losses and their impacts on freshwater, cropland, and fertiliser use. Science of the Total Environment 438, 477–489.
Kupaei, M.A., Garmakhany, A.D., 2014. Effect of microencapsulated essential oils on the storage life of mango fruit (*Mangifera indica.* L, cv Chaunsa). Minerva Biotecnologica 26, 49–55.
Lee, D.S., Yam, K.L., Piergiovanni, L., 2008. Food Packaging Science and Technology. CRC Press, Boca Raton.
Liu, J., Lundqvist, J., Weinberg, J., Gustafsson, J., 2013. Food losses and waste in China and their implication for water and land. Environmental Science and Technology 47, 10137–10144.
Liu, J., Sui, G.L., He, Y.Z., Liu, D.J., Yan, J., Liu, S.X., Qin, W., 2014. Prolonging storage time of baby ginger by using a sand-based storage medium and essential oil treatment. Journal of Food Science 79, M593–M599.
Lu, J., Charles, M.-T., Vigneault, C., Goyette, B., Raghavan, G.S.V., 2010. Effect of heat treatment uniformity on tomato ripening and chilling injury. Postharvest Biology and Technology 56 (2), 155–162.
Marsh, K., Bugusu, B., 2007. Food packaging – roles, materials, and environmental issues. Journal of Food Science 72, R39–R55.
Nahman, A., de Lange, W., 2013. Costs of food waste along the value chain: Evidence from South Africa. Waste Management 33, 2493–2500.

Nahman, A., de Lange, W., Oelofse, S., Godfrey, L., 2012. The costs of household food waste in South Africa. Waste Management 32, 2147–2153.

Oldring, P.K.T., Nehring, U., 2007. Packaging Materials 7. Metal Packaging for Foodstuffs. International Life Sciences Institute, Belgium.

Orsat, V., Raghavan, G.S.V., 2007. Use of microwave in post harvest technology of fruits and vegetables. Fresh Produce 1 (1), 16–22.

Parfitt, J., Barthel, M., Macnaughton, S., 2010. Food waste within food supply chains: quantification and potential for change to 2050. Philosophical Transactions of the Royal Society B-Biological Sciences 365, 3065–3081.

Prusky, D., 2011. Reduction of the incidence of postharvest quality losses, and future prospects. Food Security 3, 463–474.

Raghavan, G.S.V., Sosle, V., 2006. Grain drying (Chapter 23). In: Mujumdar, A.S. (Ed.), Handbook of Industrial Drying. CRC Press, pp. 563–574.

Raheem, D., 2013. Application of plastics and paper as food packaging materials - an overview. Emirates Journal of Food and Agriculture 25, 177–188.

Rhim, J.W., Park, H.M., Ha, C.S., 2013. Bio-nanocomposites for food packaging applications. Progress in Polymer Science 38, 1629–1652.

Rojas-Grau, M.A., Oms-Oliu, G., Soliva-Fortuny, R., Martin-Belloso, O., 2009. The use of packaging techniques to maintain freshness in fresh-cut fruits and vegetables: a review. International Journal of Food Science and Technology 44, 875–889.

Serrano, M., Martinezmadrid, M.C., Martinez, G., Riquelme, F., Pretel, M.T., Romojaro, F., 1996. Review: role of polyamines in chilling injury of fruit and vegetables. Food Science and Technology International 2, 195–199.

Shi, W., Dumont, M.J., 2014. Review: bio-based films from zein, keratin, pea, and rapeseed protein feedstocks. Journal of Materials Science 49, 1915–1930.

Tefera, T., Kanampiu, F., De Groote, H., Hellin, J., Mugo, S., Kimenju, S., Beyene, Y., Boddupalli, P.M., Shiferaw, B., Banziger, M., 2011. The metal silo: an effective grain storage technology for reducing post-harvest insect and pathogen losses in maize while improving smallholder farmers' food security in developing countries. Crop Protection 30, 240–245.

Thompson, A.K., 1998. Controlled Atmosphere Storage of Fruits and Vegetables. CAB International, Oxon. New York, NY.

Thompson, A.K., 2003. Fruit and Vegetables : Harvesting, Handling, and Storage. Blackwell Pub, Oxford, UK. Ames, Iowa, Iowa State Press.

Tian, W.N., Lv, Y.C., Cao, J.K., Jiang, W.B., 2014. Retention of iceberg lettuce quality by low temperature storage and postharvest application of 1-methylcyclopropene or gibberellic acid. Journal of Food Science and Technology-Mysore 51, 943–949.

Valero, D.S.M., 2010. Postharvest Biology and Technology for Preserving Fruit Quality. CRC Press, Boca Raton.

Vigneault, C., Goyette, B., Raghavan, G.S.V., 1995. Continuous flow liquid-ice system tested on broccoli. Canadian Agricultural Engineering 37, 225–230.

Index

Note: Page numbers followed by "f" "t" and "b" indicate figures, tables and boxes respectively.

FSC
www.fsc.org
MIX
Papier | Fördert gute Waldnutzung
FSC® C083411

Kolibri 360 GmbH
Ferdinand-Jühlke-Straße 7
99095 Erfurt, Deutschland
produktsicherheit@kolibri360.de